Other Kaplan Books for College-Bound Students

College Admissions and Financial Aid

Conquer the Cost of College
Parent's Guide to College Admissions
Scholarships
The Unofficial, Unbiased Insider's Guide to the 320 Most Interesting Colleges
The *Yale Daily News* Guide to Succeeding in College

Test Preparation

ACT
AP Biology
AP Calculus AB: An Apex Learning Guide
AP Chemistry: An Apex Learning Guide
AP English Language and Composition: An Apex Learning Guide
AP English Literature and Composition
AP Macroeconomics/Microeconomics: An Apex Learning Guide
AP Statistics: An Apex Learning Guide
AP U.S. Government & Politics: An Apex Learning Guide
AP U.S. History: An Apex Learning Guide
SAT & PSAT
SAT Math Mania
SAT Math Workbook
SAT Verbal Velocity
SAT Verbal Workbook
SAT II: Biology
SAT II: Chemistry
SAT II: Mathematics Levels IC and IIC
SAT II: Physics
SAT II: Spanish
SAT II: U.S. History
SAT II: Writing

AP PHYSICS B

By Connie Wells with Hugh Henderson

Simon & Schuster

SYDNEY · LONDON · SINGAPORE · NEW YORK

* AP is a registered trademark of the College Entrance Examination Board, which neither sponsors nor endorses this book.

Kaplan Publishing
Published by Simon & Schuster
1230 Avenue of the Americas
New York, New York 10020

Copyright © 2002 by Kaplan, Inc.

All rights reserved. No part of this book may be reproduced or transmitted in any form or by any means, electronic or mechanical, including photocopying, recording, or by any information storage and retrieval system, without the written permission of the Publisher, except where permitted by law.

Kaplan® is a registered trademark of Kaplan, Inc.

For all references in this book, AP is a registered trademark of the College Entrance Examination Board, which is not affiliated with this book.

For bulk sales to schools, colleges, and universities, please contact: Order Department, Simon and Schuster, 100 Front Street, Riverside, NJ 08075. Phone: (800) 223-2336. Fax: (800) 943-9831.

Contributing Editor: Seppy Basili
Project Editor: Jessica Shapiro
Interior Design and Page Layout: Martha P. Arango
Cover Design: Cheung Tai
Production Manager: Michael Shevlin
Editorial Coordinator: Déa Alessandro
Executive Editor: Del Franz

Special thanks are extended to Angelo Ferrari, Vincent Jeffrey, Scott Johns, Larissa Shmailo and Michael Wolff.

Manufactured in the United States of America
Published simultaneously in Canada

November 2002
10 9 8 7 6 5 4 3 2 1

ISBN: 0-7432-2590-2

The material in this book is up-to-date at the time of publication. However, the College Entrance Examination Board and Educational Testing Service (ETS) may have instituted changes in the test after this book was published. Be sure to carefully read the materials you receive when you register for the test.

If there are any important late-breaking developments—or any changes or corrections to the Kaplan test preparation materials in this book—we will post that information online at **kaptest.com/publishing**. Check to see if there is any information posted there regarding this book.

Table of Contents

About the Authors vi

SECTION I: THE BASICS

Chapter 1: Inside the AP Physics Exam 3

Chapter 2: Strategies for Succeeding on the Exam 11

SECTION II: PHYSICS REVIEW

Chapter 3: The Mathematical Tools for AP Physics 23

Chapter 4: The Laboratory Experience 37

Chapter 5: Newtonian Physics 41

Chapter 6: Fluid Mechanics and Thermal Physics 107

Chapter 7: Electricity and Magnetism 135

Chapter 8: Waves and Optics 195

Chapter 9: Atomic and Nuclear Physics 239

SECTION III: FULL-LENGTH PRACTICE TESTS

Practice Test I 269

Answers and Explanations for Practice Test I 287

Practice Test II 307

Answers and Explanations for Practice Test II 325

ABOUT THE AUTHORS

Connie Wells has a master of science degree in physics from the University of Kansas and is a physics and Advanced Placement Physics teacher as well as Science Department Chair at Pembroke Hill School in Kansas City, Missouri. During the past fifteen years as an A.P. Physics teacher, she has been active in test scoring and development. She is currently a Table Leader for the A.P. Physics Reading and presents teacher workshops throughout the country.

Contributing Author: Significant text for Chapters 5 through 9 was contributed by Hugh Henderson, who has taught physics at Plano Senior High School in Plano, Texas since 1985. Hugh received his bachelor of science degree in physics from Stephen F. Austin State University, and earned a master's degree in physics and teaching from the University of Texas at Dallas in 1991. He has taught AP Physics since 1988, and has lead numerous workshops in the United States and abroad for teachers and students in science curriculum and test preparation.

AUTHOR'S PREFACE

This book has been designed for your use in preparation for the Advanced Placement Physics B Examination. In it you will find a list of topics covered by the AP syllabus, information about the format of the exam, test preparation advice, strategies on approaching both the multiple choice and free response sections, concise review material on each topic with sample questions, and a set of practice problems with solutions at the end of each section. Changes in the exam incorporated in 2002 that include attention to the addition of new topics, the deletion of some topics, and modifications in the equation sheet are included. A separate section discusses how to approach the increased emphasis on laboratory questions on the examinations and includes a prototype lab. Finally, two complete 3-hour practice exams are included, with answers and explanations for all multiple choice questions and complete solutions to the free response section.

If you have completed a one or two year course in physics in preparation for the exam, this manual will provide review materials and additional practice with questions that parallel the AP format. Important to your preparation is gaining perspective on time limitations; timed practice with questions designed along AP format will give you an "edge" in transferring the physics you already know onto the answer sheet.

Even if you are not planning to take the AP Physics Examination, the text and sample problems within can be an effective supplement for any physics course.

Good luck in your pursuits!

Connie Wells

THE BASICS

Section I

Inside the AP Physics Exam

CHAPTER 1

Before you begin studying for the AP Physics B exam, let's take a step back and consider the style and format of the exam to get the larger picture. Knowing the type of exam, how it is scored, and the stated goals for the course might help in focusing on how to prepare.

THE ADVANCED PLACEMENT (AP) PROGRAM

Through the Advanced Placement Program, you can take college-level courses while you are in high school. Based on your score on an AP Exam, colleges and universities can grant you placement or college credit, or both.

In addition to getting a head start on your college coursework, you can improve your chances of acceptance to competitive schools, since colleges know that AP students are better prepared for the demands of college courses. You can also save money on tuition if you receive credit!

Structure of the AP Physics B Exam

The AP Physics B exam is administered in May by the College Board's AP Services. It is one of two examinations given simultaneously in Physics: AP Physics B and AP Physics C. The AP Physics B syllabus covers a broad range of topics, and the test itself is not calculus based. The AP Physics C syllabus has two parts—Mechanics, and Electricity and Magnetism—and is calculus based. This text will help to prepare you for the AP Physics B exam, which is three hours long and is divided into two equally weighted sections:

KAPLAN   3

AP Physics
CHAPTER 1

Section I—Multiple Choice

You will have 90 minutes to answer 70 multiple choice questions. Neither a calculator nor a formula sheet is allowed on this section, but a *Table of Information* is provided.

Section II—Free-Response

The second section of the exam consists of 6 or 7 problems, each with several sections, to solve in 90 minutes. You may use a calculator for this section, and both a *Formula Sheet* and a *Table of Information* are provided.

AP Physics B Exam at a Glance

	Number of Questions	Time (minutes)	Calculator Permitted?	Formula Sheet?
Section I (Multiple Choice)	70	90	no	no
Section II (Free Response)	6–7	90	yes	yes

Calculator Policy

You may use a calculator only during the Free Response section of the exam, which is during the second 90-minute period. A scientific calculator is recommended, and graphing capability may come in handy. Any programmable calculator is allowed as long as it does not have a typewriter-like or QWERTY keyboard. Visit the AP Physics Website at www.collegeboard.com/ap/students/physics for more information about the AP Physics examination policies.

TOPIC OUTLINE FOR AP PHYSICS B

The AP Physics B exam covers the major concepts outlined below. Percents indicate the percentage of the exam (combined multiple choice and free response) represented by that topic.

Newtonian Mechanics (35%)

Kinematics (7%)

Motion in one dimension

Motion in two dimensions

Newton's Laws of Motion (9%)

Static Equilibrium

Single Particle Dynamics

Systems of Two or More Bodies

4 **KAPLAN**

Work, Energy, and Power (5%)
    Work and Work-Energy Theorem
    Conservative Forces and Potential Energy
    Conservation of Energy
    Power

Systems of Particles and Linear Momentum (4%)
    Impulse and Momentum
    Conservation of Linear Momentum and Collisons

Circular Motion and Rotation (4%)
    Uniform Circular Motion
    Torque and Rotational Statics

Oscillations and Gravitation (6%)
    Simple Harmonic Motion (dynamics and energy relationships)
    Mass On a Spring
    Pendulum and Other Oscillations
    Newton's Law of Gravity
    Orbits of Planets and Satellites (circular only)

Fluid Mechanics and Thermal Physics (15%)

Fluid Mechanics (5%)
    Hydrostatic Pressure
    Buoyancy
    Fluid Flow Continuity
    Bernoulli's Equation

Temperature and Heat (3%)
    Mechanical Equivalent of Heat
    Specific and Latent Heat
    Heat Transfer and Thermal Expansion

Kinetic Theory and Thermodynamics (7%)
    Ideal Gases
      Kinetic Model
      Ideal Gas Law
    Laws of Thermodynamics
      First Law (including PV diagrams)
      Second Law

Electricity and Magnetism (25%)
Electrostatics (5%)
   Charge, Field, and Potential
   Coulomb's Law and Field and Potential of Point Charges
   Fields and Potentials of Planar Charge Distributions
Conductors and Capacitors (4%)
   Electrostatics with Conductors
   Parallel Plate Capacitors
Electric Circuits (7%)
   Current, Resistance, Power
   Steady-State Direct Current Circuits with Batteries and Resistors
   Steady-State Capacitors in Circuits
Magnetostatics (4%)
   Forces on Moving Charges in Magnetic Fields
   Forces on Current-Carrying Wires in Magnetic Fields
   Fields of Long Current-Carrying Wires
Electromagnetic Induction (including Faraday's and Lenz's Law) (5%)

Waves and Optics (15%)
Wave motion (5%)
   Properties of Traveling Waves
   Properties of Standing Waves
   Doppler Effect
   Superposition
Physical Optics (5%)
   Interference and Diffraction
   Dispersion of Light and the Electromagnetic Spectrum
Geometric Optics (5%)
   Reflection and Refraction
   Mirrors
   Lenses

Atomic and Nuclear Physics (10%)

Atomic Physics and Quantum Effects (7%)

Photons and the Photoelectric Effect

Atomic Energy Levels

Wave-Particle Duality

Nuclear Physics (3%)

Nuclear Reactions (with conservation of mass number and charge)

Mass-Energy Equivalence

Other Skills and Miscellaneous Topics

Vectors and Scalars

Vector Mathematics

Graphs of Functions

History of Physics

Contemporary Topics in Physics

SCORING THE AP PHYSICS EXAM

The AP Physics B exam is scored on a scale of 1 to 5, with 5 being the highest score. The scores are defined as follows:

5   Extremely well qualified

4   Well qualified

3   Qualified

2   Possibly qualified

1   No recommendation

Keep in mind that each college decides for itself which AP scores will be accepted for advanced placement or college credit. Many schools award one semester of placement or credit for a score of 3 or higher on the exam, but some schools might require a 4 or a 5. There are some schools that do not accept AP scores at all. Additionally, credit or placement may sometimes be given in an elective but not in the department in which the student is majoring. It is often a good idea to consult with the Physics Department Chair of the institution to which you are applying to clarify policies on granting of AP placement or credit. It is a good idea also, as we will discuss further, to maintain a portfolio or notebook of physics laboratory work to present for credit consideration.

The AP examinations are not actually "curved." However, some multiple-choice questions are retained from year to year and used to calibrate the difficulty level of that exam. Student performance on calibrated questions is compared to performance on the remainder of the exam in order to determine where to set the cutoff for each score level. This calibration allows standards to be set to ensure that scores reflect the same statistical strength of performance each year. Thus, the cutoff level for each score does not remain the same from year to year.

How Are Exams Graded?

The multiple-choice section of the exam is graded by computer. These multiple choice booklets are then shredded and are not available for review by the student after the examination. The free-response section is graded by faculty consultants—college professors and high school AP teachers who are specially trained to assess student performance on these questions. Students' names and schools are concealed on the free-response booklets to ensure fairness and anonymity in the grading process. Additionally, as each reader assesses a question, scores on previously scored questions are masked to prevent bias in assessment of the question. Scoring rubrics are carefully developed and applied to each question. Copies of these scoring rubrics are made available to teachers and students for training purposes. Students may order their free response booklet after the reading, since the exam readers do not make markings on the booklets during the reading. Remaining booklets are shredded after September.

Scoring the Multiple-Choice Questions

For multiple-choice questions, there is a penalty for answering incorrectly, as opposed to leaving the question blank. The base score on this section is:

Number of questions answered correctly $- \dfrac{1}{4}$ *(Number of questions answered incorrectly).*

For example, suppose a student answered 40 of the 70 multiple choice questions correctly, left 10 questions blank, and answered 20 questions incorrectly. That student's score on the multiple choice section would be calculated thus: $40 - \dfrac{1}{4}(20) = 35$.

Since the multiple-choice section counts as half the overall score on the exam, the score on the multiple choice section is multiplied by a factor 1.286 (or by $\dfrac{90}{70}$) to make the value of that section equivalent to the 90 points assigned to the free-response section. In the example above, the student has then made a multiple-choice score of 35×1.286, or 45 points, which will be added to the free-response score to obtain the test total.

Scoring the Free-Response Questions

Each free-response question is assigned a point value, usually 10 or 15 points. The faculty consultants scoring a particular question set up scoring guidelines that define carefully what answers will be accepted for each section of that question and how many points will be assigned to each section. Each question is then assigned a score from 0 to either 10 or 15, depending on the point value of the question and the quality and accuracy of answers. The total point value for each question is printed next to the question number in the booklet. The total score on the free-response section is simply the sum of the scores on each of the questions, for a maximum possible score of 90 points. The free-response total is added to the multiple-choice score to determine the final test score.

How Do I Get My Grades?

AP Grade Reports are sent in July to each student's home, high school, and any colleges designated by the student. At the time of the test, students may designate the colleges to which they would like their grades sent on the answer sheet. Students may also contact AP Services to forward their grade to other colleges after the exam, or to cancel or withhold a grade.

AP Grades By Phone

AP Grades are available by phone for $13 a call beginning in early July. A touch-tone phone and valid credit card are needed. The toll-free number is (888) 308-0013.

REGISTERING FOR THE EXAM

To register for the AP Physics exam, contact your school guidance counselor or AP Coordinator. If your school does not administer the exam, ask your guidance counselor to contact AP Services at (609) 771-7300 for test sites.

Fees

The fee for each AP Exam is $78. The College Board offers a $22 credit to qualified students with acute financial need. A portion of the exam fee may be refunded if a student does not take the test. There is a $20 late fee for late exam orders. Check with AP Services for applicable deadlines or go to the AP web site for information on exam dates: www.collegeboard.com/ap.

ADDITIONAL RESOURCES

The College Board offers a number of publications about the Advanced Placement Program, including *Advanced Placement Program Course Description–Physics, A Guide to the Advanced Placement Program*, and the *AP Bulletin for Students and Parents*. You can order these online at www.collegeboard.com/ap/ (use the store links) or call AP Services for an order form. The AP Physics page on this web address also now contains a wealth of information for teachers and students, including sample tests.

For More Information

For more information about the AP Program and or the AP Physics exams, contact your school's AP Coordinator or guidance counselor, or contact AP Services:

AP Services
P.O. Box 6671
Princeton, NJ 08541-6671
Phone: (609) 771-7300; (888) CALL-4-AP
TTY: (609) 882-4118
Fax: (609) 530-0482
E-mail: apexams@info.collegeboard.org
Website: http://www.collegeboard.com/ap

Strategies for Succeeding On the Exam

CHAPTER 2

This chapter will help you develop test taking strategies for the Advanced Placement Physics B Examination. A passing grade (3, 4, or 5) on the examination is accepted for advanced placement or credit equivalent to a semester course in physics at many colleges or universities.

To be prepared for the AP Physics exam, you should have a good grasp of physics concepts as well as proficiency in other skills necessary for physics such as: graphing and graphical analysis; use of the calculator; algebra, trigonometry, and elementary functions; vectors and scalars; and designing and analyzing experiments.

HOW TO USE THIS BOOK

Each chapter of this book lays out important objectives of the AP Physics B course, with examples to illustrate. The second half of the chapter is devoted to practice questions—sample free-response practice problems, with answers immediately following. Here's a suggestion for using these questions to best benefit: For each question or problem, use a sheet of paper to cover the solution following the problem. Work out your own solution on the paper, then slide it down to reveal the solution and discussion.

At the end of the book you'll find two complete Practice Tests with solutions. It is best to save these tests until you are ready to try them under simulated testing conditions. In other words, work through them as best you can within the time stipulated to gain experience in working at the pace necessary to complete the actual exam. Alternatively, if you cannot work an entire free-response section, just choose one problem. The point value of the problem is a recommended time limit in minutes.

KAPLAN 11

ESTIMATING YOUR SCORE

The scale used to assign A.P. grades varies from year to year, depending upon the difficulty level of the test. However, these scales are published by the College Board every five years and do not vary appreciably. After taking a sample exam under timed conditions, you can get an approximation of the score you might have been assigned using the scale below, which is a composite of several years' published results. Keep in mind that this will reflect your ability most accurately if determined from an entire two-part multiple choice and free response test taken during a timed 3-hour period. For smaller sections of the test, using methods described in the previous chapter, you might adjust your result to a 180 point total and apply it to the scale.

Sample Grading Scale

124–180	5
95–123	4
59–94	3
40–58	2
0–39	1

Strategies for Success on the Multiple Choice Section

On this section, you are given 90 minutes to complete 70 multiple choice questions. Since the score for the section is based upon number correct, with a ¼ point penalty per question answered incorrectly, the goal here is to answer as many questions correctly as possible within the time period.

To reduce the effect of the ¼ –point penalty, it is advised that you only answer questions for which you have a good idea about the concepts involved, or can narrow the answer choices to two. By narrowing the choices, you can take a "guess" on the answer with an estimated 50% chance of getting it correct. However, for questions on which you are having to spend an inordinate amount of time or have little clue what the answer could be, it is best to simply leave that question blank and move on.

The test is often arranged somewhat sequentially, with mechanics questions early in the exam, followed by waves, light, electricity, etc. However, while you may experience the level of questions becoming more difficult for you as you move through the exam, keep in mind that there are often some quite simple questions at the end of the exam. Even though your course may not have covered modern physics, for example, don't automatically assume that you cannot answer questions in this area of the exam. There may be some fairly simple, straightforward questions in this area that you know and can easily answer.

The multiple choice section requires you to answer questions without the use of a calculator. Be assured that the test is designed to be taken without the need of a calculator, even though many questions require numerical calculations. For those questions requiring numerical

The Basics

STRATEGIES FOR SUCCEEDING ON THE EXAM

solutions, use the space next to the question to write out the formula or relationship that describes the situation. Then substitute the numbers given. Look for opportunities to simplify by canceling or to estimate the answer by rounding quantities before the final calculation. The answer choices are usually selected in wide enough ranges that rounding during calculation should not hinder your ability to select the best answer. Additionally, look for other clues to the best answer, such as correct units, signs, or answers that are simply not reasonable for the situation given.

Example:

What is the force required to bring a 35 kg object to a stop from a speed of 20 m/s over a distance of 12 meters?

(A) 220 N
(B) −220 N
(C) 580 N
(D) −580 N
(E) 1000 N

Reasoning: We know that in order to stop the object, a force opposing the motion, or negative force, must be applied. Thus, only answers (B) and (D) are possible choices. Using the formula $v_f^2 = v_o^2 + 2as$ to determine the acceleration:

$$0 = (20 \text{ m/s})^2 + (2)(a)(12 \text{ m})$$

We can round the displacement to 10 m instead of 12 m to quickly estimate the acceleration to be about −20 m/s^2. Then, using $F = ma$, we estimate the force to be about −700 N, which is closer to answer (D). Since we know that we underestimated the displacement, we know also that the acceleration will be a little high and our estimation of the force a little high.

After you take the sample tests, use the *Explanations to Multiple Choice Questions* at the back of this book for insights into quick solutions to multiple choice questions.

Most questions are conceptual in nature, requiring you to compare quantities, give reasons, or demonstrate knowledge of relationships from formulas. **It is essential that you know the formulas that apply to the concepts in the Physics B syllabus**. For this reason, it will be helpful for you to obtain a copy of the equation list that will be attached to your exam. This list can be obtained from the current copy of the booklet, *Advanced Placement Course Description in Physics*, which is available from the College Board (check their website). Familiarity with these equations will be necessary in solving numerical questions, and writing down and examining the relationships between quantities in these equations will provide insights into the answers to many conceptual questions that ask for relationships between variables.

KAPLAN 13

AP Physics
CHAPTER 2

Some questions ask you to solve for one variable in terms of another variable in the question.

Sample:

A particle of charge q moves through a magnetic field of strength B at an angle θ to the direction of the field. The particle experiences a force of magnitude F. With what speed is the particle moving?

(A) $\dfrac{F}{qB}$

(B) $\dfrac{qB}{F}$

(C) $\dfrac{qB \sin \theta}{F}$

(D) $\dfrac{F}{qB \sin \theta}$

(E) qBF

Reasoning: First, we can eliminate choice (E), because the units are not reasonable. With knowledge of the formula $F = qvB \sin \theta$, we may also recognize that choices (A) and (C) simply do not have enough information. However, the most direct way to work this one is to simply write the equation next to the problem and solve for v, leading us to the correct choice (D).

It is important to note here that writing out a few simple steps in the space next to the problem in your booklet is not only allowable, it is also wise! Taking that few extra seconds to write out relationships can help to avoid simple errors leading to the wrong answer choice. **You can believe that the writers of the examination do not simply write a problem and provide you with the correct answer plus four ridiculous answers**. The exam writers have selected choices that probably include answers that could be arrived at by making simple incorrect maneuvers. Take time to make a few notes that will clarify your thinking.

You should also be familiar with the various types of format for the multiple choice questions. In all cases, there are five answers from which to choose: (A), (B), (C), (D), and (E).

- Most questions are a straightforward question or problem with five answer choices, as in Samples 1 and 2, above. In some cases, you will also be given a drawing or diagram to clarify the situation. To reduce confusion, diagrams given are almost invariably placed *above* the questions to which they apply.

- One important scenario is what we might call a "problem set," in which you are first given a drawing, diagram, or graph with a description, which we call the "stem." Below that may be 3 or 4 questions all pertaining to the stem and diagram above. *It is important that you recognize this and refer to the stem and diagram when answering all questions.* These "problem sets" are easy to recognize, as they will almost invariably be in one column on the page and there will be a heading at the top that says, for example: "Questions 31–34." Then you know that all four of those questions refer to the beginning stem and diagram. It is also important to know that *it is rare*

14 **KAPLAN**

STRATEGIES FOR SUCCEEDING ON THE EXAM

that the solutions to questions in these groups depend upon the answers to previous questions. Thus, even though you may not be able to answer one or more questions in the group, do not abandon remaining questions. There may be subsequent questions in the group that you are able to answer.

- Another type of multiple choice question is what we might call the "selection group." Your answer choice here depends on the selection of one or more choices from a group of answers given as choices I, II, or III.

Example:

Which of the following physical phenomena decrease as the inverse square of the distance?

    I. Sound Intensity
   II. Gravitational Force
  III. Electric Potential

(A) I only
(B) II only
(C) I and II
(D) II and III
(E) I, II, and III

Reasoning: The correct answer is (C), since both sound intensity and gravitational force diminish as the inverse square of distance. In the event that you are not sure about one of the choices, it may still be possible to make a reasonable answer selection, based upon choices given. For example, since we know that electric potential, $\frac{kQ}{R}$, does not follow the inverse square law, we know immediately that any answer choice with III in it is not correct, eliminating choices (D) and (E). Thus, even if we are not sure about sound intensity, we have still reduced the answer choices sufficiently to make a guess at the answer.

Strategies for Success on the Free Response Section:

On this section, you are given 90 minutes to complete 6 or 7 free response questions—six at 15 points or four at 15 points plus three at 10 points. The goal in this section is to complete as much as possible in each of the problems given. Here, you are supplied with both a pink copy and a green copy of the questions, each containing an *information sheet* and an *equation sheet*. You are also allowed to use a calculator on this section. (See calculator policies, above.)

First, as you organize quickly to begin work, *set aside the green copy of the test, using it as a reference for diagram or opening it to the Equation Sheet and Information Sheet for reference as you work.* Only the pink copy is sent in to be graded, so time spent writing on the green sheet is really not recommended. Spend your time on the pink copy of the test.

AP Physics
CHAPTER 2

Second, quickly glance through the test to get an idea of how many problems you have to work and approximately how much time you can spend on each problem. The point value of each problem should be used as an approximate guide to the time allotted to work on that problem. A 15-point problem, for example, should be worked in about 15 minutes.

The free response problems are generally organized along the order of the syllabus, with mechanics first, then waves and optics, electricity and magnetism, and modern physics. However, keep in mind that these problems are often multi-conceptual, with topics mixed within problems. Often, the first problem on the test will be a straightforward motion problem that will take less than the allotted time, leaving you a little extra time for more complex problems later on.

As you work each problem, it is important to remember the following pointers:

- Read the introductory paragraph, or "stem", carefully before you begin work. It contains all necessary background information, describes the situation (which may also be described in a diagram), and sets out some instructions for the solution.

- Show all your work clearly in the space provided. In the past, readers have been required to look through all pages provided for that problem for possible solutions, *but this is no longer the case.* Readers are required to grade only what the student writes in the space provided, *unless the student leads the reader to another space by writing "see bottom of next page" and then clearly labels the part continued or draws an arrow to the continuation of the solution.*

- Do not leave extraneous formulas or duplicate solutions, but do not take time to erase. Simply cross out anything you do not want graded. In the past, readers have been required to select the correct or "best" solution from what the student has written, even if it is has been crossed out. *This is no longer the case.* Readers are usually required to grade whatever comes first, even if a correct solution is provided subsequently. Readers will not grade anything that is crossed out, even if it is correct. Additionally, students who play the game of writing down a multitude of formulas from the equation sheet, hoping that at least one is correct, will "lose the game." Readers are instructed to grade only what the student sets out as the solution; extraneous formulas will detract from that solution, so that even the correct formula(s) may not be given credit.

- In each solution, clearly show the formula(s) used, substituting numerical values *with units* where applicable. Credit is given for partial solutions, and partial credit may be given for incorrect solutions when work is clearly shown. Additionally, *the correct answer may be given no credit at all if no work is shown to justify how the solution was obtained.* Though it may seem time-consuming to write all this work in detail, it will pay off in the long run, as it is easy to make simple errors under the stress of a timed test. Showing intermediate steps may help you to accumulate significant credit, even for an incorrect solution.

- At least one free response problem will generally be of the "symbolic" type, where only symbols and no numerical values are given. On these, answers must be given using only the variable symbols given and fundamental constants, such as g and π. It

16 KAPLAN

The Basics

STRATEGIES FOR SUCCEEDING ON THE EXAM

is very important to make all substitutions in the final answer so that the answer is given only using symbols defined in the stem. The next example below addresses this.

- Most free response questions are set up in sequence so that, for example, part (a) leads you to part (b), and so on. However, there are times when you may not know how to do an early section, where the answer is used as part of the solution in a subsequent section. The trick here is to either assign an arbitrary value to the section you were not able to work, then show clearly how that would be substituted later, *or* when you set up your solution in the later section, to simply show that unknown answer straightforwardly.

Example:

An object of mass m is being spun in a vertical circle at the end of a string of length l. The maximum tension the string can withstand without breaking is T_{max}. In terms of the given values and fundamental constants, find:

(A) The minimum speed at the top of the circle at which the mass must be spun to take a circular path

(B) The tension in the string at the bottom of the circle if the mass maintains this speed

(C) The maximum speed, v_{max}, at which the mass may be spun without breaking the string

(D) The subsequent motion of the mass if the string breaks at the bottom of the circle when traveling at speed v_{max}

Reasoning: (a) Using $F_{net} = \dfrac{mv^2}{R}$, with no tension in the string and the weight of the object, mg, providing the centripetal force at the minimum speed:

$$mg = \frac{mv^2}{l} \text{ and } v = \sqrt{gl}$$

Note that we are able to use g in the answer, since it is a fundamental constant, but we must substitute l for the radius, since the length of the string is given as a variable, and R is not a given symbol.

(b) Now, at the bottom of the circle, $F_{net} = T - mg$ and $F_{net} = \dfrac{mv^2}{R}$

Then: $T - mg = \dfrac{mv^2}{R}$ and $T = mg + \dfrac{mv^2}{R}$

However, we cannot leave the answer in this form and expect full credit, since neither v nor R were given to us in the stem as a variable. Thus, we must substitute l for R and our answer from part (a) for v. The correct answer would be:

$$T = mg + \frac{m(gl)}{l}$$

Or, better: $T = 2mg$

KAPLAN 17

AP Physics
CHAPTER 2

In this sample, if you did not have the correct answer to part (a), but you had clearly shown the substitution of whatever you got to (a) into part (b), you would receive <u>full credit</u> for part (b), even though your final answer to part (b) is incorrect. Another possibility is that you did not work part (a) at all but could see how to set up part (b). You might show your solution to part (b) thus, which would also receive full credit:

$$T - mg = m\frac{(answer\ to\ part\ a)^2}{R}$$

$$T = mg + m\frac{(answer\ to\ part\ a)^2}{l}$$

Exam readers often see test questions that receive 14/15 with no correct final answers on a question of several parts, when a simple error in part (a) carries through all the parts of the question.

Let's look at more hints for working out problems on the test:

- With the recent emphasis on evaluation of laboratory work, you can expect one question that is laboratory or exploration-based. These are often purposely designed to be unique to students, whether or not they have done extensive laboratory work in their preparatory course. Students who have done extensive laboratory work should have an advantage here, as they are familiar with equipment and have developed the ability to design and analyze experiments. However, *even though you may not have done laboratory work in your course, you may be able to answer portions, if not all, of this question, based upon your knowledge of physics.* Many students make the mistake of skipping this question entirely, believing they will be wasting their time on it since they are not familiar with the situation. You may receive significant credit by attempting this problem, or portions of it. See Section II, Chapter 4 in this book for more clarification on the laboratory-based question.

- Though time is limited and you may expect to leave some portions of the free response section incomplete, it is to your benefit to make an attempt at every problem. Credit given to these questions is sometimes "front-loaded", that is, a significant portion of the point credit is given to early sections of the problem. You may find the later sections of some problems to get significantly more difficult and time-consuming. Thus, it may be to your benefit, particularly if time is running short, to go on to a problem you have not yet attempted, rather than leaving it completely blank. The time it might take for the last, most difficult section of one problem might be enough to make it through several early sections of another problem. Additionally, even a problem on a subject that you have not covered or do not know well may have sections that are quite straightforward and could be solved using your general knowledge of physics.

- Take a ruler or straight edge in with you to use when constructing "best fit" line graphs or when making ray diagrams on optics problems. When a ray or line is required, the line is often checked with a straight edge by the reader to confirm that it is really meant to be a line and not a curve.

The Basics
STRATEGIES FOR SUCCEEDING ON THE EXAM

- When drawing rays for optics problems, clearly show each ray with arrows on it to show the direction of light path. It is generally advisable to show actual light paths with solid lines and arrows and projected light paths (in drawing virtual images) with dotted or dashed lines and arrows.

- When you are asked to construct or label a "free body diagram," make sure each force has an arrow on it and is clearly labeled. Do not show both a force and its components; this would be redundant and would essentially "double" that force. *Make sure you show only forces on free body diagrams; do not show velocity or acceleration vectors, for example.*

- Whenever a graph is given, immediately ask yourself the significance of the slope and/or the area under the curve. These concepts are frequently asked for directly or become an important part of the solution.

- When a problem asks for an answer and then says "justify your answer" or "explain your reasoning," be clear but succinct. A justification or explanation could be either conceptual or mathematical, as long as there is enough to provide a physical basis for your answer. Avoid the temptation to provide a lengthy explanation, as it is not necessary and you could fall into the trap of saying too much—actually providing erroneous "facts" that could negate your explanation.

- "Speak to the reader." Remember that these exams are graded by physicists who will understand, or try to understand, what you relate to them on paper. Be clear but concise, and do as much as you can on each problem that relates to the question.

- Some problems ask you to describe or draw the subsequent **motion** or **path** of an object. These two words have quite different meanings. In describing the subsequent path of an object, you need only to describe the direction or change in direction the object takes. The motion of an object implies not only the path or trajectory but any changes in velocity or acceleration.

Example: (referring back to the previous example) In part (d), you are asked for the subsequent motion if the string breaks at the bottom. A correct answer might be: *The mass will initially move horizontally, tangent to the circular path, at speed* v_{max}, *continuing at that speed horizontally as it accelerates from an initial vertical velocity of zero with an acceleration, g.* This answer is sufficient to get full credit. Saying that the mass leaves the circular path on a tangent at speed v_{max} and describes a parabolic path to the ground would probably also get full credit, since the parabolic path implies acceleration in one direction and constant speed in the other direction.

If this problem had asked for a description of the <u>path</u> of the object instead, it would have been correct to either draw a parabolic path, with the object moving initially horizontally on a tangent from the circle, or to say: *The object moves off on a tangent horizontally as it falls vertically to the ground in a parabolic path.* The challenge here is to describe the path sufficiently so there is no ambiguity about the shape. Drawing might be the best option.

A few more tips:

- If you change your mind about an answer, clearly cross out the portion you do not want to have graded—don't waste time and energy erasing. However, *do not cross out until you have new information to provide.* It's amazing how often students cross out

KAPLAN   19

correct answers, then don't provide anything else. In those cases, readers are usually required to give no credit—even though the correct answer is visible! Even if you believe your solution is incorrect, it may have portions that are correct and that will receive partial credit in the scoring.

- Make sure you answer the question(s) asked in each part. Sometimes students get involved in the calculation and forget to answer one part of the question or give an answer that is not asked!

- During the solution of a problem, always carry one or two extra digits in the calculation. *Do not round to the correct number of significant figures until the final answer.* Excessive rounding during the calculation will throw off your result. However, avoid the temptation to carry too many digits in your final answer. Final answers on the A.P. Physics B Exam are generally given full credit if they show the number of significant figures within one or two of the accepted number of digits.

Preparing for Exam Day

Once you have completed a course and have reviewed sufficiently, you should be ready to take the AP Physics B Examination. Avoid the temptation to spend long hours just prior to the exam day "cramming" for the test. You should remember that this test is designed to assess your background in Physics B topics and your readiness to move on to more advanced work or to receive credit for your previous work. It is a lengthy, rather exhausting test, so your best preparation just prior to exam day is rest. Make sure you have gathered the materials you will need for the test (calculator, pencils, a small clock or watch, and a straight edge or ruler). Any time you have to spend on review during the day prior to the test might be best spent simply going over a copy of the Equation Sheet or becoming familiar with a copy of the Information Sheet that will be provided with the test. The real key to success on this exam is advance preparation and the confidence to go in on exam day and relate as much as you can about your knowledge of physics.

PHYSICS REVIEW

Section II

The Mathematical Tools for AP Physics

CHAPTER 3

FACT AND FORMULA REVIEW

Quadratic Formula:

When a quadratic equation is written in standard form $ax^2 + bx + c = 0$,

$$x = \frac{-b \pm \sqrt{b^2 - 4ac}}{2a}$$

Multiplying Binomials (FOIL):

Multiply the First terms, the Outer terms, the Inner terms, then the Last terms. To factor, do the opposite.

For example: $(2x + 3)(3x + 4) = 6x^2 + 8x + 9x + 12$, or $6x^2 + 17x + 12$

Slope of a Line: $\frac{\Delta y}{\Delta x}$ or rise/run

Basic Geometry Facts:

- When two parallel lines are intersected by a transversal, the following angles are equal in measure:

    Alternate interior
    [e.g., $c = b$ and $g = f$]
    Corresponding
    [e.g., $f = b$ and $g = c$]

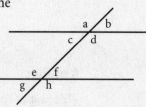

- Vertical angles are always equal in measure.
- The sum of the measures of the angles of a triangle is 180°.
- Area of a rectangle: $A = bh$
- Area of a triangle: $A = \frac{1}{2}bh$

AP Physics
CHAPTER 3

- Area of a circle: $A = \pi r^2$
- Circumference of a circle: $C = 2\pi r$
- Volume of a cylinder: $V = \pi r^2 h$
- Surface area of a cylinder: $A = 2\pi r^2 + 2\pi r h$
- Volume of a sphere: $V = \dfrac{4}{3}\pi r^3$
- Surface area of a sphere: $A = 4\pi r^2$
- Volume of parallelepiped: $V = lwh$

Right Triangle Trigonometry:

Pythagorean Theorem:

$a^2 + b^2 = c^2$

SOH-CAH-TOA

$\sin = \dfrac{\text{opp}}{\text{hyp}}$

$\cos = \dfrac{\text{adj}}{\text{hyp}}$

$\tan = \dfrac{\text{opp}}{\text{adj}}$

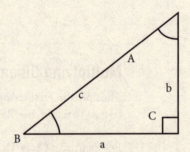

$\sin B = \dfrac{b}{c}$

$\cos B = \dfrac{a}{c}$

$\tan B = \dfrac{b}{a}$

Graphs:

This plot of "y versus x" is a direct, linear relationship of the form $y = mx$, where m is the slope.

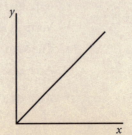

This plot of "y versus x" is a direct, linear relationship of the form $y = mx + b$, where m is the slope and b is the y-intercept.

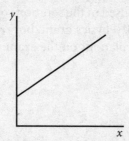

This plot of "y versus x" is a root curve of the form $y = k\sqrt{x}$. The plot of "y versus square root of x" will produce a line with slope k.

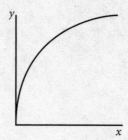

This plot of "y versus x" is a power curve of the form $y = kx^2$. The plot of "y versus the square of x" for the same data will produce a line with slope k.

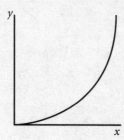

This plot of "y versus x" is an inverse curve of the form $y = \dfrac{k}{x}$. The plot of "y versus the reciprocal of if x" for the same set of values will produce a line with slope k.

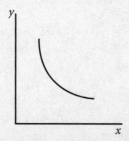

AP Physics

CHAPTER 3

PRACTICE MATH ASSESSMENT

This practice test is designed to "jog your memory" regarding mathematical and trigonometric methods sometimes used in the solutions of problems. These are not generally of the type found on the actual A.P. Physics exam, but rather they are meant to assess your readiness to perform necessary calculations on the exam.

1. Solve for x:

   $6(x + 1) = 14x - 6$

   (A) 0
   (B) 0.625
   (C) 0.875
   (D) 1.5
   (E) 1.14

2. Solve for x:

   $x^2 = 6x + 16$

   (A) $x = 0, 4$
   (B) $x = 0, -0.25$
   (C) $x = -2, 8$
   (D) $x = 0$
   (E) No real solutions

3. Solve for x:

   $2x^2 - x = 0$

   (A) $x = -1, -3$
   (B) $x = -1.73, 1.73$
   (C) $x = -.95, 1.45$
   (D) $x = 0, 0.5$
   (E) No real solutions

4. Solve for t:

   $2t^2 - 6t = 0$

   (A) $t = 0, 3$
   (B) $t = 0, -1$
   (C) $t = 3$
   (D) $t = 0$
   (E) Not possible

26 **KAPLAN**

5. Solve for r in the equation $F = \dfrac{mv^2}{r}$.

   (A) $\dfrac{F}{m}$

   (B) $\dfrac{mv^2}{F}$

   (C) Fmv^2

   (D) $\dfrac{F}{mv^2}$

   (E) Fvm

6. The number of micrometers in one centimeter is:
   (A) 10^{-6}
   (B) 10^{-4}
   (C) 10^2
   (D) 10^4
   (E) 10^6

7. For triangle ABC (shown below, not drawn to scale), fill in the blank:

   $\sin A =$

   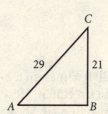

   (A) $\dfrac{29}{21}$

   (B) $\dfrac{21}{29}$

   (C) 1.05
   (D) 8
   (E) Not enough information given

8. Again using triangle ABC from problem #7, find the base length AB.
   (A) 8
   (B) 20
   (C) 16
   (D) 2.8
   (E) 64

9. A flag pole that is 8 meters tall casts a shadow on the ground that extends 6 meters along the ground from the base of the flag pole. What is the distance from the top of the flag pole to the end of the shadow?

   (A) 3 meters
   (B) 4 meters
   (C) 5 meters
   (D) 6 meters
   (E) 10 meters

10. Referring to problem #9, what angle does the imaginary line from the tip of the shadow to the top of the flag pole make with the ground?

    (A) 30°
    (B) 37°
    (C) 45°
    (D) 53°
    (E) 60°

11. Which of the following functions is equivalent to the one shown?

    $y = \cos(90° - x)$

    (A) $y = \sin(x)$
    (B) $y = \sin(90° - x)$
    (C) $y = \cos x$
    (D) $y = \cos(x + 180°)$
    (E) $y = \cos(x + 90°)$

12. A streetlight is 6 meters tall. It shines on a 2-meter-tall post standing 3 meters from the base of the streetlight. How long is the shadow of the post? This picture, not drawn to scale, shows the situation.

    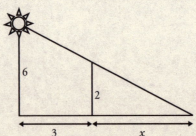

    (A) 6 meters
    (B) 3 meters
    (C) 2 meters
    (D) 1.5 meters
    (E) Not enough information

THE MATHEMATICAL TOOLS FOR AP PHYSICS

13. Which of the following is the closest approximation of 60 feet per second in meters per second?

(A) 18 m/s
(B) 30 m/s
(C) 60 m/s
(D) 100 m/s
(E) 180 m/s

14. A plane leaves the ground at an air speed of 450 km/h and rises at an angle of 40°. At what speed does the plane appear to be moving along the ground surface?

(A) 675 km/h
(B) 605 km/h
(C) 500 km/h
(D) 345 km/h
(E) 300 km/h

15. Find the slope of the line $x = 2$.

(A) 1
(B) 2
(C) 0; this line is horizontal.
(D) The slope is undefined; this line is vertical.
(E) $\dfrac{1}{2}$

16. For which of the following functions is the value at $t = 0$ equal to 0?

    I. $y = \sin(t)$
   II. $y = \cos(t)$
  III. $y = -\sin(t)$

(A) I only
(B) I and II
(C) I and III
(D) II only
(E) II and III only

17. The graph of the equation $y = -16t^2 - \cos(20°)t + 300$ is:

(A) a line.
(B) a cosine curve.
(C) a parabola
(D) a circle.
(E) a root curve.

AP Physics
CHAPTER 3

18. Which of the following are equivalent forms of the same SI unit?

    (A) meter, millisecond
    (B) kilogram, pound
    (C) mile, inch
    (D) centimeter, kilometer
    (E) gram, ounce

19. Which of the following, according to its order of magnitude, is approximately the same size as a human child?

    (A) 10^{-3} m
    (B) 10^{-2} m
    (C) 10^1 m
    (D) 10^3 m
    (E) 10^0 m

20. Which of the following is a unit that exists both in British and SI systems?

    (A) meter
    (B) second
    (C) pound
    (D) gram
    (E) foot

21. If light travels at 186,000 miles per second, how far, in meters, is a light year (the distance light travels in one year)?

    (A) 3×10^8 meters
    (B) 5.87×10^{12} meters
    (C) 9.46×10^{15} meters
    (D) 6.79×10^7 meters
    (E) 186,000 meters

THE MATHEMATICAL TOOLS FOR AP PHYSICS

22. The following variables are listed with their units:

Variable	Unit
x	meters
F	kilogram·meter/second^2
v	meters/sec
t	second
m	kilogram
a	meters/second^2

In which of the following equations are the units <u>not</u> balanced?

(A) $x = vt$
(B) $Ft = mv$
(C) $x = at^3$
(D) $F = ma$
(E) $v = at$

23. While driving in Europe, you see a sign indicating that the speed limit is 120. What are the most likely units for this speed?

(A) kilometers/second
(B) meters/second
(C) miles/second
(D) miles/hour
(E) kilometers/hour

24. Which of the following measurements has the fewest significant figures?

(A) 0.00001 cm
(B) 12.6 meters
(C) 101 kg
(D) 11.534 seconds
(E) 54 m/s

25. How many significant figures are in 1.004?

(A) 1
(B) 2
(C) 3
(D) 4
(E) 5

AP Physics
CHAPTER 3

Multiple Choice Answers:

1. D
2. C
3. D
4. A
5. B
6. D
7. B
8. B
9. E
10. D
11. A
12. D
13. A
14. D
15. D
16. C
17. C
18. D
19. E
20. B
21. C
22. C
23. E
24. A
25. D

THE MATHEMATICAL TOOLS FOR AP PHYSICS

FREE–RESPONSE QUESTIONS:

1. A child stands before a tall building and wants to know how high it is. She notices that it is a sunny day, so armed with a meter stick, the fact that she is 1.2 meters tall, and a knowledge of similar triangles, she proceeds to measure the building's height by comparing the length of her shadow to the length of the building's shadow.

   (a) Draw a diagram that depicts the situation in the problem and label those quantities the child can directly measure herself. (At this point you don't need to write numerical values.)

Answer:

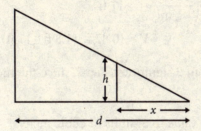

h = child's height
x = length of child's shadow
d = length of building's shadow

   (b) Assume the child is 1.2 meters tall and casts a shadow of 6.4 meters. Determine the angle of the sun at that moment.

Answer:

$$h = 1.2 \text{ m and } x = 6.4 \text{ m}$$

$$\tan(\text{sun's angle with the ground}) = \frac{h}{x}$$

$$\text{angle} = \tan^{-1}\left(\frac{1.2}{6.4}\right) = 10.5 \text{ degrees}$$

   (c) If the building casts a shadow of 77.9 meters, how tall is it?

Answer:

Since corresponding sides of similar triangles are proportional,

$$\frac{\text{height of building}}{d} = \frac{\text{height of child}}{x}$$

$$\frac{\text{building height}}{77.9 \text{ m}} = \frac{1.2 \text{ m}}{6.4 \text{ m}}$$

$$\text{building height} = 14.6 \text{ m}$$

AP Physics
CHAPTER 3

2. A rectangular slab of concrete is cut from a large block. The slab measures 10.4 inches thick, 1.8 feet long, and 0.0021 miles wide.

   (a) Find its volume in cubic feet.

Answer:

$$V = l \times w \times h$$

First, convert all measurements to feet.

$$(10.4 \text{ in})\left(\frac{1 \text{ ft}}{12 \text{ in}}\right) = 0.87 \text{ ft}$$

$$(0.0021 \text{ mi})\left(\frac{5280 \text{ ft}}{1 \text{ mi}}\right) = 11 \text{ ft}$$

$$V = (0.87 \text{ ft})(1.8 \text{ ft})(11 \text{ ft}) = 17 \text{ ft}^3$$

The final answer is rounded to 2 significant digits, since the measurement 1.8 ft has only 2 significant digits.

   (b) Convert this value to SI units of cubic meters.

Answer:

$$(17 \text{ ft}^3)\left(\frac{1 \text{ m}}{3.28 \text{ ft}}\right)^3 = 0.48 \text{ m}^3$$

3. A wheelchair ramp is to be rated at 8%, which means that it rises 8 cm for every 100 cm of horizontal distance. The height of the ramp is 35 cm.

   (a) What is the length of this ramp?

Answer:

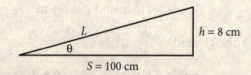

By definition, the height of any ramp of this type must be 8% of its horizontal span. Thus, a ramp with 35 cm height must have a horizontal span, S, such that 8% of S is equal to 35 cm:

$$(0.08)(S) = 35 \text{ cm}$$

$$S = \frac{35 \text{ cm}}{0.08} = 438 \text{ cm}$$

Now that we know the horizontal and vertical components of the ramp, use the Pythagorean Theorem to find the ramp length, L:

$$(35 \text{ cm})^2 + (438 \text{ cm})^2 = L^2$$

$$L = 440 \text{ cm}$$

Physics Review

THE MATHEMATICAL TOOLS FOR AP PHYSICS

(b) What is the angle this ramp makes with the horizontal?

Answer:

The angle is the same for both triangles, so that the tangent of the angle is height divided by horizontal span:

$$\tan \theta = \frac{8 \text{ cm}}{100 \text{ cm}} \text{ or } \frac{35 \text{ cm}}{438 \text{ cm}}$$

$$\theta = 4.5°$$

(c) How far would the incline have to be extended (minimum length) to meet government regulations of a maximum 3% incline?

Answer:

For a 3% incline, the height, 35 cm, needs to be 3% of the horizontal span, S:

$$35 \text{ cm} = (0.03)(S)$$

$$S = \frac{35 \text{ cm}}{0.03} = 1{,}167 \text{ cm}$$

Use the Pythagorean Theorem with the height and horizontal span to find the length of the ramp, L:

$$(35 \text{ cm})^2 + (1{,}167 \text{ cm})^2 = L^2$$

$$L = 1{,}167.5 \text{ , or } 1{,}200 \text{ cm}$$

4. List the three most common base units and at least one additional unit that may be derived from them.

Answer:

In the SI (or metric) system, the three base units are the kilogram (a unit of mass), meter (a unit of length), and second (a unit of time). The Newton, a unit of force, is equal to one kilogram-meter per second squared.

5. Convert a speed limit of 70 miles per hour to the SI unit of meters/second.

Answer:

$$(70 \text{ mi/h})(1{,}609 \text{ m/1 mi})(1 \text{ h/3{,}600 s}) = 31 \text{ m/s}$$

6. Show that the Impulse-Momentum Theorem produces units that balance.

$F\Delta t = m\Delta v$, where F = force (Newtons or kg·m/s^2), t = time (second), m = mass (kilograms), and Δv = change in velocity (meters/second).

Answer:

Simply show that the substituted units are the same on both sides of the equation.

$$(kg\text{–}m/s^2)(s) = (kg)(m/s)$$

KAPLAN 35

The Laboratory Experience

CHAPTER 4

In the past several years, the AP Physics Examination has increased its assessment of the laboratory experience of AP Physics students by adding a laboratory question to the free response section. Since specific experiments are not prescribed in the syllabus, each of these questions is designed to allow the student to demonstrate knowledge of proper experimental techniques without requiring performance of a specific experiment. However, wide experience in the physics laboratory certainly increases the student's proficiency in setting up equipment and analyzing data. Students enrolled in a regular course containing laboratory work should keep a notebook or report portfolio. These, in conjunction with results from the AP Physics B Exam, can be presented for laboratory credit or placement in some colleges or universities.

LABORATORY SKILLS

From laboratory experience, the student should demonstrate the ability to:

- **Design an experiment**, given a purpose or problem to solve, or design an experiment from an original idea.
- **Present data in a clear fashion**. Often the most concise method of presenting data is in a data table, with symbols used for variables and units on all measurements.
- **Analyze data to a solution of the problem or verification of the hypothesis.** Data analysis may be as simple as averaging data, but a more common method employs the graphing of a "best fit" curve, then determining the relationship between variables and/or the meaning of the slope or area under the curve.
- **Discuss and analyze errors within the experiment.**
- **Evaluate experimental design and/or make suggestions for other experiments related to the purpose.**

SAMPLE EXPERIMENT

The following simple experiment may serve as an example for students who have not had extensive experience in the laboratory or have not had a large role in experimental design:

The teacher has assigned the problem of determining a value for g, the acceleration due to gravity, using common laboratory equipment. The student is to design and perform the experiment. Any number of methods might be employed, such as: timing a simple pendulum, timing an oscillating spring, or examining the motion of a falling object. Computer-interfaced devices are available to some students that may provide printed graphs and data for the student to examine. However, students are encouraged to use simple equipment, as the results can often be just as accurate.

Teachers may vary in their requirements for style in the experiment report, but these basic elements should be included. On the AP Physics B Exam, experimental questions are usually divided into parts, leading the student to each of the following categories:

Purpose: To determine a value for g, the acceleration due to gravity, using a simple pendulum.

Materials Needed: string, mass or pendulum bob, ring stand, clamp, extension bar clamp, stopwatch

Background: The pendulum equation $T = 2\pi \sqrt{\dfrac{l}{g}}$ shows the relationship between the two

variables, length and period. The accepted value for "g" is 9.8 m/s^2. Since neither angle of release nor mass appear in the equation, their values will not affect the result. However, using an angle of release less than 15 degrees will assure that the pendulum is in harmonic motion, and using a larger mass (around 100 g) will give the pendulum more inertia, further reducing the effect of additional vibrations. Extending the pendulum over the edge of the table allows use of longer lengths, producing longer periods of oscillation and minimizing the effect of human reaction time in timing. Length of the pendulum should be measured from the point of attachment of the string to the stand to the center of mass of the pendulum bob.

Method: Clamp the ring stand base firmly to the edge of a table to reduce additional vibrations. From the ring stand post, attach an extension rod so that the pendulum can suspend over the edge of the table. Attach the mass or pendulum bob to the string and attach the string to the extension rod so the pendulum hangs over the edge of the table. Pull the pendulum bob back to an angle of 10–15 degrees and release it. Use the stopwatch to time 20 complete oscillations of the pendulum and divide the total time by the number of oscillations to find the period of the pendulum. Take three different sets of timings for each of three different pendulum lengths.

Physics Review
THE LABORATORY EXPERIENCE

Data:

Trial	Length, l (m)	Number of Swings	Total Time (s)	Period, T (s)
1	0	0	0	0
2	1.00	20	40.4	2.02
3	1.25	20	45.1	2.25
4	1.50	20	48.2	2.40

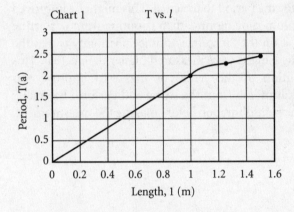
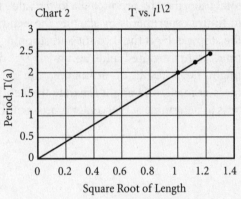

Data Analysis: Average period was determined (see data table) for each of three lengths. Note that the stopwatch was readable only to the nearest 0.1 second, but the number of swings was exactly 20, so dividing time by number of swings produced period data with three significant digits.

A plot of "Period versus Length" (see Chart 1 above) showed that period is proportional to the square root of length, so a second plot of "Period versus Square Root of Length" (see Chart 2 above) was produced, and the best fit line was drawn between the data points. The slope of this line is 2.0 s/m$^{1/2}$, and the proportionality constant between T and $l^{1/2}$ is the slope. Using the pendulum equation above, $T = 2\pi \left(\dfrac{l}{g} \right)^{1/2}$, this proportionality constant must be $2\pi/g^{1/2}$. Setting the slope and the proportionality constant equal to each other and solving for g:

$$\frac{2\pi}{g^{\frac{1}{2}}} = 2.0$$

$$g = \left(\frac{2\pi}{2.0} \right)^2 \text{ and } g = 9.9 \text{ m/s}^2$$

AP Physics

CHAPTER 4

This value can then be compared to the accepted value for g, 9.8 m/s^2, by calculating percentage error:

$$\% \text{ Error} = \frac{\text{accepted value} - \text{experimental value}}{\text{accepted value}} \times 100\%$$

$$\% \text{ Error} = \frac{9.8 - 9.9}{9.8} \times 100\% = 1.0\%$$

Discussion: The value of 9.9 m/s^2 for g is 1.0% higher than the accepted value. Assuming an accurate measurement of the lengths, the value for period must have been measured lower than actual. In the timing process, a slow start with the stopwatch might have produced a lower total time, thereby producing a lower value for the period than actual. This might be improved in future experiments by starting the pendulum moving first, then counting from 0, starting the stopwatch on the 0 count and stopping it on the 20 count. Another possibility is that the times were measured quite accurately but the lengths were measured higher than actual. This might be a result of estimating the position of the center of mass. To improve this measurement, one might separate the pendulum bob from the string and perform balancing tests to determine actual center of mass of the pendulum bob before making the measurement.

40   KAPLAN

Newtonian Physics

CHAPTER 5

I. KINEMATICS

Kinematics is the study of *how* things move—how far (*distance* and *displacement*), how fast (*speed* and *velocity*), and how fast the aforementioned *how fast* changes (*acceleration*). We say that an object moving in a straight line is moving in *one dimension*, and an object moving in a curved path (like a *projectile*) is moving in *two dimensions*. We relate all these quantities with a set of equations called the *kinematic equations*.

Key Terms

Acceleration: the rate of change in velocity

acceleration due to gravity: the acceleration of a freely falling object in the absence of air resistance, which near the earth's surface is approximately 10 m/s^2

acceleration-time graph: plot of the acceleration of an object as a function of time

average acceleration: the acceleration of an object measured over a time interval

average velocity: the velocity of an object measured over a time interval; the total change in displacement divided by the total time taken to change the displacement

constant (or uniform) acceleration: acceleration which does not change during a time interval

constant (or uniform) velocity: velocity which does not change during a time interval

KAPLAN **41**

AP Physics

CHAPTER 5

displacement: change in position in a particular direction (vector)

distance: the length moved between two points (scalar)

free fall: motion under the influence of gravity

initial velocity: the velocity at which an object starts at the beginning of a time interval

instantaneous: the value of a quantity at a particular instant of time, such as instantaneous position, velocity, or acceleration

kinematics: the study of how motion occurs, including distance, displacement, speed, velocity, acceleration, and time

position: the distance between an object and a reference point

position-time graph: the graph of the motion of an object that shows how its position varies with time

projectile: any object that is projected by a force and continues to move by its own inertia

range of a projectile: the horizontal distance between the launch point of a projectile and where it returns to its launch height

reference point: zero location in a coordinate system

speed: the ratio of distance to time (scalar)

trajectory: the path followed by a projectile

velocity: ratio of the displacement of an object to a time interval (vector)

velocity-time graph: plot of the velocity of an object as a function of time, the slope of which is acceleration

Distance and Displacement

Distance can be defined as total length moved. If you run around a circular track you have covered a distance equal to the circumference of the track. Distance is a scalar, which means it has no direction associated with it. *Displacement*, however, is a vector. Displacement is defined as the straight-line distance between two points, and is a vector which points from an object's initial position toward its final position. In our previous example, if you run around a circular track and end up at the same place you started, your displacement is zero, since there is no distance between your starting point and your ending point.

42 **KAPLAN**

Speed and Velocity

Average speed is defined as the amount of distance a moving object covers divided by the amount of time it takes to cover that distance:

$$\text{average speed} = v = \frac{\text{total distance}}{\text{total time}} = \frac{d}{t}$$

where v stands for speed, d is for distance, and t is time.

Average velocity is defined a little differently than *average speed*. While average speed is the total change in distance divided by the total change in time, average velocity is the *displacement* divided by the change in time. Since velocity is a vector, we must define it in terms of another vector, displacement. Often, average speed and average velocity are interchangeable for the purposes of the AP Physics B exam. Speed is the magnitude of velocity; that is, speed is a scalar and velocity is a vector. For example, if you are driving west at 60 miles per hour, we say that your speed is 60 mph, and your velocity is 60 mph west. We will use the letter v for both speed and velocity in our calculations, and will take the direction of velocity into account when necessary.

Acceleration

Acceleration a is defined as the rate of change of velocity, or the change in velocity divided by the change in time t:

$$\text{acceleration} = a = \frac{\Delta v}{\Delta t}$$

Acceleration is a vector that tells us how fast velocity is changing and in what direction. For example, if you start from rest on the goal line of a football field and begin walking up to a speed of 2 m/s for the first second, then up to 4 m/s, for the second second, then up to 6 m/s for the third second, you are speeding up with an average acceleration of 2 m/s for each second you are walking. We write:

$$a = \frac{\Delta v}{\Delta t} = \frac{2 \text{ m/s}}{1 \text{ s}} = 2 \text{ m/s/s} = 2 \, \frac{\text{m}}{\text{s}^2}$$

In other words, you are changing your speed by 2 m/s for each second you walk. If you start with a high velocity and slow down, you are still accelerating, but your acceleration would be considered negative, compared to the positive acceleration discussed above.

Usually, the change in speed Δv is calculated by the final speed v_f minus the initial speed v_i. The initial and final speeds are called *instantaneous* speeds, since they each occur at a particular instant in time and are not average speeds.

The Kinematic Equations for Uniformly Accelerated Motion

Kinematics is the study of the relationships between distance and displacement, speed and velocity, acceleration, and time. The kinematic equations are the equations of motion which relate these quantities to each other. These equations assume that the acceleration of an object is *uniform*, that is, constant for the time interval we are interested in. The kinematic equations listed below would not work for calculating velocities and displacements for an object which is accelerating erratically. Fortunately, the AP Physics B exam generally deals with uniform acceleration, so the kinematic equations will be very helpful in solving problems on the test. Vector quantities are in bold print.

Kinematic Equations

$$\bar{v} = \frac{d}{t}$$

$$d = \frac{1}{2}(v_f + v_o)t$$

$$\bar{a} = \frac{\Delta v}{\Delta t} = \frac{v_f - v_o}{\Delta t}$$

$$v_f = v_o + at$$

$$s = v_o t + \frac{1}{2}at^2$$

$$v_f^2 = v_o^2 + 2as$$

where t = time; d = distance traveled; v_i = initial velocity; v_f = final velocity, or the velocity at the end of a time interval; a = acceleration; s = displacement (or change in position, $\Delta x = x_f - x_i$).

Notice that in some of the equations above we have chosen to replace the time interval Δt with a simple t, and the distance interval Δd with a simple d. Also, these equations are the scalar form of the kinematic equations rather than the vector form. We need to remember that if we have a moving object with an acceleration in the opposite direction of its velocity, we must give either the velocity or the acceleration a negative sign to distinguish direction.

When solving problems using the kinematic equations, we need to write down which quantities are given in the statement of the problem and the quantity we want to find, and use the equation that includes all of them.

Free Fall

An object is in *free fall* if it is falling freely under the influence of gravity. Any object, regardless of its mass, falls near the surface of the Earth with an acceleration of 9.8 m/s^2, which we will denote with the letter g. We will round the free fall acceleration g to 10 m/s^2 for the purpose of the AP Physics B exam. This free fall acceleration assumes that there is no air resistance to impede the motion of the falling object, and this is a safe assumption on the AP Physics B test unless you are told differently for a particular question on the exam.

Since the free fall acceleration is constant, we may use the kinematic equations to solve problems involving free fall. We simply need to replace the acceleration a with the specific free fall acceleration g in each equation. Remember, any time a velocity and acceleration are in opposite directions (like when a ball is rising after being thrown upward), you must give one of them a negative sign.

Kinematic Equations	Free Fall Equations
$v_f = v_o + at$	$v_f = v_o + gt$
$s = v_o t + \frac{1}{2} a t^2$	$s = v_o t + \frac{1}{2} g t^2$
$v_f^2 = v_o^2 + 2as$	$v_f^2 = v_o^2 + 2gs$

Projectile Motion

Projectile motion results when an object is thrown either horizontally through the air or at an angle relative to the ground. In both cases, the object moves through the air with a constant horizontal velocity, and at the same time is falling freely under the influence of gravity. In other words, the projected object is moving horizontally and vertically at the same time, and the resulting path of the projectile, called the *trajectory*, has a parabolic shape. For this reason, projectile motion is considered to be *two-dimensional* motion.

The motion of a projectile can be broken down into constant velocity and zero acceleration in the horizontal direction, and a changing vertical velocity due to the acceleration of gravity. Let's label any quantity in the horizontal direction with the subscript x, and any quantity in the vertical direction with the subscript y. If we fire a cannonball from a cannon on the ground pointing up at an angle θ, the ball will follow a parabolic path and we can draw the vectors associated with the motion at each point along the path:

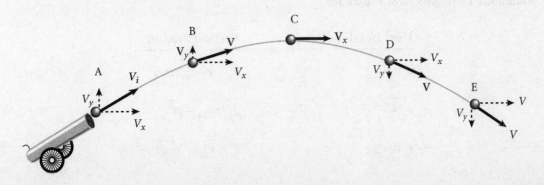

At each point, we can draw the horizontal velocity vector v_x, the vertical velocity vector v_y, and the vertical acceleration vector g, which is simply the acceleration due to gravity. Notice that the length of the horizontal velocity and the acceleration due to gravity vectors do not change, since they are constant. The vertical velocity decreases as the ball rises and increases as the ball falls. The motion of the ball is symmetric; that is, the velocities and acceleration of the ball on the way up is the same as on the way down, with the vertical velocity being zero at point C and reversing its direction at this point.

At any point along the trajectory, the velocity vector is the vector sum of the horizontal and vertical velocity vectors, that is, $v = v_x + v_y$.

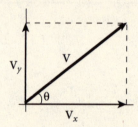

By the Pythagorean theorem,

$$v = \sqrt{v_x^2 + v_y^2}$$

and

$$v_x = v \cos \theta$$

$$v_y = v \sin \theta$$

$$\theta = \tan^{-1} \left(\frac{v_y}{v_x} \right)$$

In both the horizontal and vertical cases, the acceleration is constant, being zero in the horizontal direction and 10 m/s² downward in the vertical direction. Therefore we can use the kinematic equations to describe the motion of a projectile.

Kinematic Equations for a Projectile

Horizontal motion	Vertical motion
$a_x = 0$	$a_y = g = 10$ m/s²
$v_x = \dfrac{s_x}{t}$	$v_y = v_{oy} - gt$
$s_x = v_x t$	$s_y = v_{oy}t - \dfrac{1}{2}gt^2$

Notice the minus sign in the equations in the right column. Since the acceleration **g** and the initial vertical velocity v_{oy} are in opposite directions, we must give one of them a negative sign, and here we've chosen to make *g* negative. Remember, the horizontal velocity of a projectile is constant, but the vertical velocity is changed by gravity.

Graphs of Motion

Let's take some time to review how we interpret the motion of an object when we are given the information about it in graphical form. On the AP Physics B exam, you will need to be able to interpret three types of graphs: displacement vs. time, velocity vs. time, and acceleration vs. time.

Displacement vs. Time

Consider the displacement vs. time graph below:

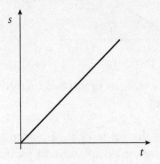

Since the graph is a straight, diagonal line, the slope is constant. The slope of a line is the rise divided by the run. The rise in this case is the displacement, or change in position, and the run is the change in time. Thus,

$$\text{slope of an } s \text{ vs. } t \text{ graph} = \frac{\text{rise}}{\text{run}} = \frac{\Delta x}{\Delta t} = \text{velocity}$$

A constant slope in this case means a constant velocity. Remember, displacement and time are changing, but the velocity (slope) is constant.

Velocity vs. Time

Consider the velocity vs. time graph below:

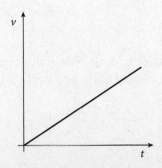

In this case, the straight diagonal line means that the velocity is changing with time, that is, the object is accelerating.

$$\text{slope of a } v \text{ vs. } t \text{ graph} = \frac{rise}{run} = \frac{\Delta v}{\Delta t} = acceleration$$

The constant slope in this case means constant acceleration. Remember, the velocity is changing with time, but the acceleration (slope) is constant.

The area under a velocity vs. time graph is the displacement.

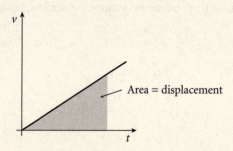

Remember, the slope of a velocity vs. time graph is acceleration, and the area under a velocity vs. time graph is the displacement.

If the velocity is changing, then the slope of a displacement vs. time graph must also be changing. For constant acceleration, the displacement vs. time graph would be a parabola:

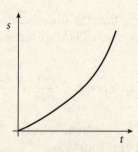

On this graph, the slope (velocity) is constantly increasing, indicating that the object is accelerating. The shape is parabolic because displacement is proportional to the square of time, as we've seen in the kinematic equations relating displacement and time for constant acceleration.

Acceleration vs. Time

Since the AP Physics B exam typically deals with only constant acceleration, any graph of acceleration vs. time on the exam would likely be a straight horizontal line:

Physics Review
NEWTONIAN PHYSICS

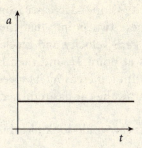

This graph tells us that the acceleration of this object is positive. If the object were accelerating negatively, the horizontal line would be below the time axis:

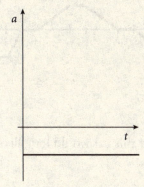

The area under either graph above would be the change in velocity during that time interval.

Consider the displacement vs. time graph below representing the motion of a car. Assume that all accelerations of the car are constant.

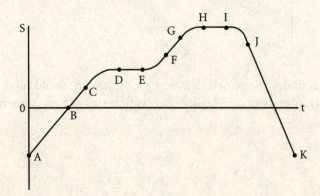

The car starts out at a position behind our reference point of zero, indicated on the graph as a negative displacement. The velocity (slope) of the car is initially positive and constant from points A to C, with the car crossing the reference point at B. Between points C and D, the car goes from a high positive velocity (slope) to a low velocity, eventually coming to rest ($v = 0$) at point D. At point E the car accelerates positively from rest up to a positive constant velocity from points *F* to *G*. Then the velocity (slope) decreases from points *G* to *H*, indicating the car is slowing down. It is between these two points that the car's velocity is positive, but its acceleration is negative, since the car's velocity and acceleration are in opposite directions. The car once again comes to rest at point *H*, and then begins gaining a negative velocity (moving backward) from rest at point *I*, increasing its speed negatively to a constant negative velocity between points *J* and *K*. At *K*, the car has returned to its original starting position.

The velocity vs. time graph for this car would look like this:

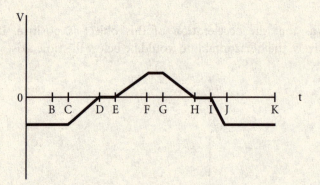

The acceleration vs. time graph for this car would look like this:

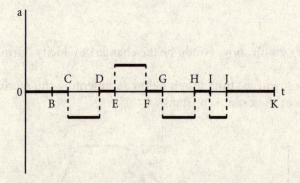

Notice that the areas under the velocity vs. time graph, which would be displacements, add up to zero. This corresponds to our earlier observation that the car returned to its starting position.

NEWTONIAN PHYSICS

Sample Problem:

An athlete wants to kick a football over the goal post from a distance of 36 meters. The ball leaves the athlete's foot with a speed of 22 m/s at an angle of 52° above the ground. The crossbar on the goal post is 3.05 meters high.

(a) How long does the ball take to reach the goal post?

Answer:

$$v_x = 22 \cos 52° = 13.55 \text{ m/s}$$

To travel a horizontal distance of 36 meters (assuming no air friction and constant velocity), using $t = \dfrac{s_x}{v_x}$, it will take

$$\frac{36 \text{ m}}{13.55 \text{ m/s}} = 2.66 \text{ s}$$

(b) What is the ball's speed when it reaches the goal post?

Answer:

Horizontal velocity is still 13.55 m/s.

Vertical velocity begins at $v_{iy} = 22 \sin 52° = 17.34$ m/s.

Use $v_{fy} = v_{iy} + at$; the final vertical velocity is:

$$17.34 \text{ m/s} + (-9.8 \text{ m/s}^2)(2.66 \text{ s}) = -8.7 \text{ m/s}$$

Combining these two components of the resultant velocity using the Pythagorean Theorem: $13.55^2 + (-8.7)^2 = v^2$.

$$v = 16.1 \text{ m/s}$$

(c) By how much is the ball above or below the crossbar when it reaches the plane of the goal post?

Answer:

The height of the ball is determined by $y_f = y_1 + v_{yi} \, t - \left(\dfrac{1}{2}\right)gt^2$.

Thus the height at $t = 2.66$ seconds is:

$$Y_f = 0 + (17.34 \text{ m/s}) - \frac{1}{2}(9.8 \text{ m/s}^2)(2.66 \text{ s})^2 = 11.45 \text{ m}.$$

Since the crossbar is only 3.05 meters above the ground, the ball goes over it by 8.4 meters.

II. NEWTON'S LAWS OF MOTION

Dynamics is the study of the *causes* of motion, in particular, *forces*. A force is a push or a pull. We arrange our knowledge of forces into three laws formulated by Isaac Newton in 1687: *the law of inertia, the law of force and acceleration* ($F_{net} = ma$), and *the law of action and reaction*.

Key Terms

air resistance: the force the air exerts on an object opposing its motion as the object moves

coefficient of friction: the ratio of the frictional force acting on an object to the normal force exerted by the surface with which the object is in contact; can be static or kinetic

dynamics: the study of the causes of motion (forces)

equilibrant: the vector which can balance a resultant vector; the force which can put a system in equilibrium

equilibrium: the condition in which there is no unbalanced force acting on a system, that is, the vector sum of the forces acting on the system is zero

force: any influence that tends to accelerate an object; a push or a pull

friction: the force that acts to resist the relative motion between two rough surfaces which are in contact with each other

inertia: the property of an object which causes it to remain in its state of rest or motion at a constant velocity; mass is a measure of inertia

inertial reference frame: a reference frame which is at rest or moving with a constant velocity; Newton's laws are valid within any inertial reference frame

kilogram: the fundamental unit of mass

kinetic friction: the frictional force acting between two surfaces which are in contact and moving relative to each other

mass: a measure of the amount of substance in an object and thus its inertia; the ratio of the net force acting on an accelerating object to its acceleration

net force: the vector sum of the forces acting on an object

newton: the SI unit for force equal to the force needed to accelerate one kilogram of mass by one meter per second squared

Newton's first law of motion: the law of inertia

Newton's second law of motion: a net force acting on a mass causes that mass to accelerate in the direction of the force

Newton's third law of motion: for every action force there is an equal and opposite reaction force

non-inertial reference frame: a reference frame which is accelerating; Newton's laws are not valid within a non-inertial reference frame

normal: a line which is perpendicular to a surface

normal force: the reaction force of a surface acting on an object

resultant: the vector sum of two or more vectors

static friction: the resistive force that opposes the start of motion between two surfaces in contact

terminal velocity: the constant velocity of a falling object when the force of air resistance equals the object's weight

weight: the gravitational force acting on a mass

Newton's First Law of Motion

The first law of motion states that an object in a state of constant velocity (including zero velocity) will continue in that state unless acted upon by an unbalanced force. Thus, a book at rest on your desk will remain at rest until you or something else applies a force to it. The property of the book which causes it to follow Newton's first law of motion is its *inertia*. Inertia is the sluggishness of an object to changing its state of motion or state of rest. It is difficult to push a car to get it moving from a state of rest because of its sluggishness or inertia, just as it is difficult to stop a car which is moving because of its sluggishness or inertia. We measure inertia by measuring the mass of an object, or the amount of material it contains. We often refer to Newton's first law as the *law of inertia*. Remember, you don't need an unbalanced force to keep an object at rest or moving at a constant velocity; its own inertia takes care of that.

Newton's Second Law of Motion

The law of inertia tells us what happens to an object when there are no unbalanced forces acting on it. Newton's second law tells us what happens to an object which *does* have an unbalanced force acting on it: it accelerates in the direction of the unbalanced force. Another name for an unbalanced force is a *net force*, meaning a force which is not canceled by any other force acting on the object. Sometimes the net force acting on an object is called an *external force*.

Newton's second law can be stated like this: *A net force acting on a mass causes that mass to accelerate in the direction of the net force. The acceleration is proportional to the force* (if you double the force, you double the amount of acceleration), *and inversely proportional to the mass of the object being accelerated* (twice as big a mass will only be accelerated half as much by the same force). In equation form, we write Newton's second law as

$$F_{net} = ma$$

where F_{net} and **a** are vectors pointing in the same direction. We see from this equation that the newton is defined as a $\frac{(kg)m}{s^2}$, and happens to be a little less than a quarter of a pound. If two ropes pull on a cart, the net force acting on the cart depends on the directions the ropes are being pulled. If the forces are in the same direction, the tensions in the ropes can be added:

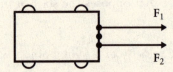

If the ropes are pulled in opposite directions, the net force is the difference between the tensions in the ropes:

If the ropes are pulled at an angle other than 0° or 180° relative to each other, we must resolve the forces into their components to find the net force:

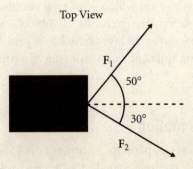

Resolving the forces into components:

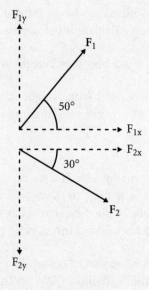

$$F_{1x} = F_1 \cos 50$$

$$F_{1y} = F_1 \sin 50$$

and

$$F_{2x} = F_2 \cos 30$$

$$F_{2y} = F_2 \sin 30$$

The sum of the x-components of the forces is the x-component of the net force:

$$F_{net\,x} = F_{1x} + F_{2x}$$

Similarly, the sum of the y-components of the forces is the y-component of the net force:

$$F_{net\,y} = F_{1y} + F_{2y}$$

By the Pythagorean theorem, the magnitude of the net force is

$$F_{net} = \sqrt{(F_{net\,x})^2 + (F_{net\,y})^2}$$

and the angle from the x-axis is $\theta = \tan^{-1}\left[\dfrac{F_{net\,y}}{F_{net\,x}}\right]$.

Weight

As we mentioned earlier, mass is a measure of the inertia of an object. If you take a 1 kg mass to the moon, it will still have a mass of 1 kg. Mass does not depend on gravity, but weight does. The *weight* of an object is defined as the amount of gravitational force acting on its mass. Since weight is a force, we can calculate it using Newton's second law:

$$F_{net} = ma \text{ becomes } \textbf{Weight} = mg$$

where the specific acceleration associated with weight is, not surprisingly, the acceleration due to gravity. Like any force, the SI unit for weight is the newton.

Static Equilibrium

A system is said to be *static* if it has no velocity and no acceleration. In other words, it's at rest and it's going to stay that way. According to Newton's first law, *if an object is in static equilibrium, the net force on the object must be zero.* That doesn't mean there are no forces acting on it; it means there are no *unbalanced* forces acting on it.

For example, three ropes are attached as shown below. The tension forces in the ropes are $\textbf{T}_1$, $\textbf{T}_2$, and $\textbf{T}_3$, and the mass of the hanging ball is m. We can find the tension in each of the three ropes by finding the vector sum of the tensions and the force of gravity on the ball and setting this sum equal to zero.

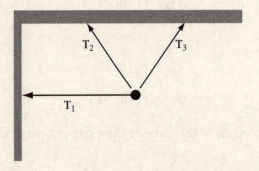

Since the system is in equilibrium, the net force on the system must be zero. We start by drawing a *free-body force diagram* of all the forces acting on the ball. A free-body force diagram is a vector diagram of all of the forces acting on an object, treating the object as if it were all located at a single point. You should always draw all of the forces acting on an object before answering a question about the forces or net force.

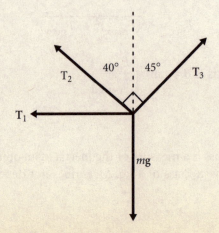

Since the forces are in equilibrium, the vector sum of the forces in the x-direction must equal zero. Using only the x-components of forces:

$$\Sigma F_x = 0$$

$$T_1 + T_{2x} = T_{3x}$$

Equation 1: $T_1 + T_2 \sin 40° = T_3 \sin 45°$

Likewise, the sum of the forces in the y-direction must also be zero:

$$\Sigma F_y = 0$$

$$T_{2y} + T_{3y} = mg$$

Equation 2: $T_2 \cos 40° + T_3 \cos 45° = mg$

Now, substitute known values and solve Equations 1 and 2 simultaneously to find the unknown tensions.

Blocks and Pulley

A typical example used to illustrate weight and Newton's second law is a blocks and pulley system. Let's look at a couple of examples of the blocks and pulley system.

In the diagram below, two blocks of mass m and $4m$ are connected by a string that passes over a pulley of negligible mass and friction. Let's find the acceleration of the system in terms of the masses and the acceleration due to gravity g.

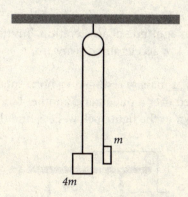

First, we should draw a free-body force diagram for each block. There are two forces acting on each of the masses: weight downward and the tension in the string upward. Our free-body force diagrams should look like this:

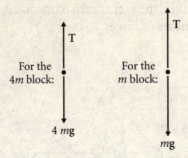

Write Newton's second law for each of the blocks, recognizing that the *4m* block will fall downward, pulling the *mg* block upward. For consistency make the direction of motion the positive direction in each equation:

$$F_{net} = 4ma \qquad F_{net} = ma$$

$$(4mg - T) = 4ma \qquad (T - mg) = ma$$

$$T = 4mg - 4ma \qquad T = mg + ma$$

The tension acting on each block is the same, and the magnitudes of their accelerations are the same. Setting their tensions equal to each other, we get:

$$mg + ma = 4mg - 4ma$$

Canceling the masses and solving for *a*, we get:

$$a = \frac{3g}{5}$$

The two blocks have the same magnitude of acceleration, but the block of mass *m* accelerates upward and the block of mass 4 *m* accelerates downward.

Let's say we have another block of mass *m* resting on a horizontal table of negligible friction. A string is tied to the block, passed over a pulley, and another block of mass 5 *m* is hung on the other end of the string, as shown in the figure below. Let's find the acceleration of the system.

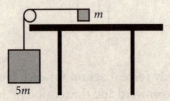

Once again, draw a free-body force diagram for each of the blocks, and then apply Newton's second law.

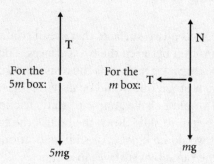

For motion in the *y*-direction only: For motion in the *x*-direction only:

$$F_{net} = 5\,ma$$
$$5\,mg - T = 5\,ma$$
$$T = 5\,mg - 5\,ma$$

$$F_{net} = ma$$
$$T = ma$$

Setting the two equations for *T* equal to each other:

$$ma = 5\,mg - 5\,ma$$
$$a = \frac{5g}{6}$$

Note that this is a reasonable answer. We expect the system to move at an acceleration less than *g*. *Always check to see that your answer is reasonable.*

Newton's Third Law

Newton's third law is sometimes called the *law of action and reaction*. It states that for every action force, there is an equal and opposite reaction force. For example, let's say your physics book weighs 6 N. If you set it on a level table, the book exerts 6 N of force on the table. By Newton's third law, the table must exert 6 N back up on the book. If the table could not return the 6 N of force on the book, the book would sink into the table. We call the force the table exerts on the book the *normal force*. Normal is another word for perpendicular, because the normal force always acts perpendicularly to the surface which is applying the force, in this case, the table. The force the book exerts on the table and the force the table exerts on the book are called an *action-reaction pair*. Action-reaction pairs act on different bodies, so they do not cancel each other.

It is important to note here that Newton's laws are useful for solving problems from the viewpoint of an *inertial reference frame*, that is, one that is at rest or moving with a constant velocity. If you are moving in a *non-inertial reference frame* (one which is accelerating), your measurements using Newton's laws will not be valid, as you may have to invent fictitious forces to describe the motions you observe as a result of relative acceleration between you and the reference frame you are trying to measure. Centrifugal force is an example of a fictitious force, which we will discuss further in a later section.

AP Physics
CHAPTER 5

Friction

Friction is a resistive force between two surfaces that are in contact with each other. Friction always opposes the relative motion between the two surfaces. There are two types of friction: *static friction* and *kinetic friction*. Static friction is the resistive force between two surfaces that are not moving relative to each other, but would be moving if there were no friction. A block at rest on an inclined board would be an example of static friction acting between the block and the board. If the block began to slide down the board, the friction between the surfaces would no longer be static. It would be kinetic, or sliding, friction. Sliding friction is typically less than static friction for the same two surfaces in contact.

We can define a ratio that gives an indication of the roughness between two surfaces in contact with each other. This ratio, called the *coefficient of friction*, is the frictional force between the surfaces divided by the normal force acting on the surfaces. The coefficient of friction is represented by the Greek letter μ (mu). Equations for the coefficients of static and kinetic friction are $\mu_s = \frac{f_{s\,max}}{N}$ and $\mu_k = \frac{f_k}{N}$ where f_s is the static frictional force and f_k is the kinetic frictional force. Note that the coefficient of static friction is equal to the ratio of the *maximum* frictional force and the normal force. The static frictional force will only be as high as it has to be to keep a system (like a block on an inclined plane) in equilibrium. When you draw a free body diagram of forces acting on an object or system of objects, you want to include the frictional force as opposing the relative motion (or potential for relative motion) of the two surfaces in contact.

For example, consider the block on the inclined plane below:

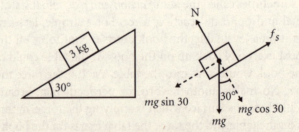

If the surface on which the block rests is not horizontal but inclined at an angle, the normal force is not equal and opposite the weight of the block. The coefficient of static friction between the block and the plane is 0.2, and the angle of incline is 30°. Will the block slide down the plane, or remain at rest? (sin 30 = cos 60 = 0.5; cos 30 = sin 60 = 0.87)

We see in the free-body force diagram that the frictional force points up the incline to oppose the motion (or impending motion) of the block. What we really want to know is whether or not the component of the weight pointing down the incline ($mg \sin 30$) is greater than the frictional force f_s pointing up the incline. Calculating each of those forces:

$mg \sin 30 = (3 \text{ kg})(10 \text{ m/s}^2)(0.5) = 15$ N down the incline

The normal force **N** is equal and opposite to the component of the weight $mg\cos 30$.

So, $f_s = \mu_s N = \mu_s(mg\cos 30) = (0.2)(3 \text{ kg})(10 \text{ m/s}^2)(0.87) = 5.2$ N up the incline.

Thus, the force down the incline is greater than the maximum frictional force up the incline, and the block will overcome friction to slide down the incline.

If the angle of incline were lower, both the component of the weight pointing down the incline ($mg \sin \theta$) and the frictional force up the incline would be less. The angle at which $mg \sin \theta$ and the frictional force are equal and opposite is called the *angle of repose*. Any angle above the angle of repose will cause the block to slide down the incline.

Air resistance is another example of a frictional force acting on a falling object or a projectile in flight. The force of air resistance is directed opposite to the direction of the motion of the object. When the air resistance becomes equal to the weight of a falling object, the object stops accelerating, but continues to fall at a constant velocity called *terminal velocity*. For a skydiver in free fall, terminal velocity is about 125 miles per hour.

Sample Problem:

Two blocks are stacked on a table. The upper block has a mass of 8.2 kg, and the lower block has a mass of 35 kg. The coefficient of kinetic friction between the lower block and the table is 0.22. The coefficient of static friction between the upper and lower blocks is 0.14. How much force can be applied horizontally to the lower block without having the upper block slide off?

[*Hint: The force of static friction between the two blocks is what makes the upper block accelerate with the lower one. To make the top block slide off, you have to accelerate the lower block so much that you exceed its maximum force of static friction.*]

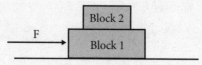

(a) Draw a free-body diagram of each of the two blocks.

Answer:

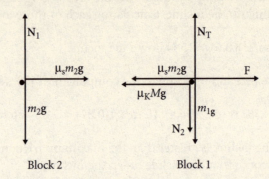

Block 2　　　　　Block 1

Where: N_1 = normal force from block 1 on block 2

$\mu_s m_2 g = \mu_s N_1$ = friction force between blocks

$m_2 g$ = weight of block 2

N_T = normal force of table on block 1

F = pushing force on bottom block

$\mu_s m_2 g$ = friction force between blocks

$\mu_K M g$ = friction force between block 1 and table, where M is total mass of both blocks

N_2 = normal force of block 2 on block 1, due to weight of block 2

$m_1 g$ = weight of block 1

(b) Show that the static forces on the overall system balance.

Answer:

$$\Sigma F_x = 0$$

$$\mu_s m_2 g - \mu_s m_2 g + F - \mu_K M g = 0$$

$$\Sigma F_y = 0$$

$$N_1 + N_T - m_2 g - N_2 - m_1 g = 0$$

(c) Write the equation relating the maximum acceleration of block 2 to the maximum force of friction between the two blocks.

Answer:

$m_2 a = \mu_s m_2 g$, where μ_s is the friction coefficient between the two blocks.

That means: $a = \mu_s\, g$.

> (d) Write the equation for the force of kinetic friction between the lower block and the table.

Answer:

The force of kinetic friction is:

$F_k = \mu_k N_T$, where N_T is the normal force from the table acting on block 1. The normal force has to balance the weight of **both** blocks and is therefore:

$$N_T = (m_1 + m_2)g$$

This yields for the kinetic friction force:

$$F_k = \mu_k(m_1 + m_2)g$$

> (e) Now, write Newton's second law that relates the acceleration of the system to the sum of all forces and calculate the maximum pushing force, using the value of the acceleration calculated in part *a*.

Answer:

Newton's second law says:

$$F_{push} - F_k = (m_1 + m_2)a$$

The two forces are the pushing force we want to calculate and the force of kinetic friction from part *d*.

Now, insert the results for the kinetic force from part *d* and for the maximum acceleration from part *c*. You get:

$$F_{push} - \mu_k\,(m_1 + m_2)g = (m_1 + m_2)\,\mu_s g$$

$$F_{push} = (\mu_k + \mu_s)\,(m_1 + m_2)\,g$$

$$= (0.22 + 0.14)\,(35\ \text{kg} + 8.2\ \text{kg})\,(9.81\ \text{m/s}^2)$$

$$= 152\ \text{N}$$

III. WORK, ENERGY, AND POWER

Work is the scalar product of the *force* acting on an object and the *displacement* through which it acts. When work is done on or by a system, the energy of that system is always changed. If work is done slowly, we say that the *power* level is low. If work is done quickly, the power level is high. If the energy of a system remains constant throughout a process, we say that energy is *conserved*.

Key Terms

conserved properties: any properties which remain constant during a process

energy: the non-material quantity which is the ability to do work on a system

joule: the unit for work or energy; equal to one Newton-meter

kinetic energy: the energy a mass has by virtue of its motion

law of conservation of energy: the total energy of a system remains constant during a process

mechanical energy: the sum of the potential and kinetic energies in a system

potential energy: the energy an object has because of its position

power: the rate at which work is done or energy is dissipated

watt: the SI unit for power equal to one joule of energy per second

work: the scalar product of force and displacement

Work and Energy

As you know by now, there are many words in physics that have well-defined meanings but are not necessarily used in the way they are normally used outside of the context of a physics course. The concept of *work* is certainly one of these words. In physics, work is defined as the scalar product of force and displacement, that is:

$$W = \mathbf{F} \cdot \mathbf{s}$$

Here, the force and displacement vectors are multiplied together in such a way that the product yields a scalar. Thus, work is not a vector, and has no direction associated with it. Since work is the product of force and displacement, it has units of newton-meters, or *joules* (J). A joule is the work done by applying a force of one newton through a displacement of one meter. One joule is about the amount of work you do in lifting your calculator to a height of one meter.

Graphically, the work done on an object or system is equal to the area under a *force* vs. *displacement* graph:

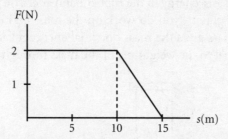

The area under the graph in this case is $(2\text{ N})(10\text{ m}) + \frac{1}{2}(2\text{ N})(15\text{ m} - 10\text{ m}) = 25$ J. Thus the force represented by the graph does 25 J of work.

In the equation for work, the scalar product of force and displacement can also be written:

$$W = Fs \cos \theta$$

where θ is the angle between the applied force and the displacement. Suppose a force **F** is applied to a box directed at 60° above the horizontal as shown in the figure below.

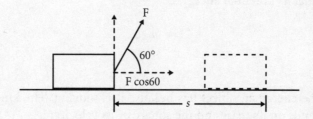

If the box is pulled by this force through a horizontal displacement *s*, the work done on the box is the horizontal component of the force (*F cos 60*) times the magnitude of the horizontal displacement *s*:

$$W = (F \cos 60)s = Fs \cos 60$$

For work to be done on an object, the force must have a component in the same direction as the displacement.

Remember, work is not a measure of how tired you are after performing the work. It is a measure of the product of the force that was applied in the direction of the displacement. Work is also a measure of the *energy* which was transferred while the force was being applied.

Energy is the ability to do work. When work is done there is always a transfer of energy. Energy can take on many forms, such as potential energy, kinetic energy, and heat energy. The unit for energy is the same as the unit for work, the *joule*. This is because *the amount of work done on a system is exactly equal to the change in energy of the system*. This is called the *work-energy theorem*. Let's look at two examples of this theorem.

Potential Energy

Potential energy is the energy a system has because of its position or configuration. When you stretch a rubber band, you store energy in the rubber band as elastic potential energy. When you lift a mass upward against gravity, you do work on the mass and therefore change its energy. The work you do on the mass gives the mass potential energy relative to the ground. To lift it, you must apply a force equal to the weight mg of the mass through a displacement height h.

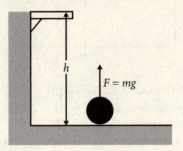

The work done in lifting the mass is

$$W = \mathbf{F} \cdot \mathbf{s} = (mg)h$$

which must also equal its potential energy, U:

$$U = mgh$$

Kinetic Energy

Kinetic energy is the energy an object has because it is moving. The kinetic energy K of a moving object depends on its mass and the square of its velocity:

$$K = \frac{1}{2}mv^2$$

In order for a mass to gain kinetic energy, work must be done on it either to push it up to a certain speed or to slow it down. The work-energy theorem states that the change in kinetic energy of an object is exactly equal to the work done on it, assuming there is no change in the object's potential energy.

$$W = \mathbf{F} \cdot \mathbf{s} = \Delta KE = \frac{1}{2}mv_f^2 - \frac{1}{2}mv_i^2$$

From the rearranging of this equation we get the familiar kinematic equation:

$$v_f^2 = v_i^2 + 2as$$

The work done on a system can also be converted into heat energy (and usually some of the work is).

Conservation of Energy

When work is done on a system, the energy of that system changes from one form to another, but the *total amount* of energy remains the same. We say that total energy is *conserved*, that is, it remains constant during any process. This is also called the *law of conservation of energy*.

As an example, let's consider a 2 kg brick that sits just on the edge of the roof of a building which is 10 meters above the ground.

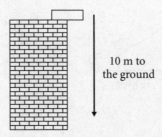

The potential energy of the brick relative to the ground is:

$$U = mgh = (2 \text{ kg})(10 \text{ m/s}^2)(10 \text{ m}) = 200 \text{ J}$$

If the brick slips off the edge of the building and falls, the potential energy of the brick begins changing into kinetic energy as the brick gains speed while losing height above the ground. Both the brick's potential and kinetic energies change, but the total amount of energy is conserved. In other words, the sum of the potential energy and kinetic energy at any time is equal to the total energy, which remains constant throughout the fall.

We can use the fact that the sum of the potential and kinetic energies remains constant during free fall to solve for quantities such as speed or initial height.

$$U_{top} + K_{top} = U_{bottom} + K_{bottom}$$

$$mgh_{top} + \frac{1}{2}mv_{top}^2 = mgh_{bottom} + \frac{1}{2}mv_{bottom}^2$$

The speed of the brick when it hits the ground can now be determined:

First, cancel the mass in each term (reinforcing the idea here that the speed of the object will be the same when it hits the ground, regardless of its mass, assuming no air friction):

$$gh_{top} + \frac{1}{2}v_{top}^2 = gh_{bottom} + \frac{1}{2}v_{bottom}^2$$

We can also assume no kinetic energy at the top and no potential energy at the bottom, so:

$$gh_{top} = \frac{1}{2}v_{bottom}^2$$

$$v_{bottom} = 14 \text{ m/s}$$

The sum of the kinetic and potential energies of a system is called the *total mechanical energy* of the system. These same principles can be applied to a block sliding down a frictionless ramp, a pendulum swinging from a height, and many other situations. We could use Newton's laws and kinematics to solve these types of problems, but usually conservation of energy is easier to apply.

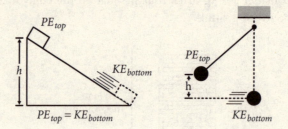

Power

Work can be done slowly or quickly, but the time taken to perform the work doesn't affect the amount of work which is done; there is no element of time in the definition for work. However, if you do the work quickly, you are operating at a higher power level than if you do the work slowly. Power is defined as *the rate at which work is done*. Often we think of electricity when we think of power, but it can be applied to mechanical work and energy as easily as it is applied to electrical energy. The equation for power is

$$P = \frac{Work}{time}$$

and has units of joules/second or *watts* (W). A machine is producing one watt of power if it is doing one joule of work every second. A 100-watt light bulb uses 100 joules of energy each second.

For example, if a motor lifts a 100 kg crate onto a deck which is 5 meters high in a time of 10 seconds, the power the machine produces is:

$$P = \frac{W}{t} = \frac{mgh}{t} = \frac{(100 \text{ kg})(10 \text{ m/s}^2)(5 \text{ m})}{10\text{s}} = 500 \text{ watts}$$

Any motor could do this same amount of work, but not necessarily as quickly, and therefore not at this 500 watt power level. For example, a small motor could lift the 100 kg crate if it were first broken into 1 kg pieces. The small motor could lift one piece at a time to the height of 5 meters and accomplish the same amount of work at a lower power level.

Sample Problem:

A boy climbs to the roof of his house and drops a 0.40 kg rubber ball from the 7.5-meter peak of the house.

(a)   How fast will the ball be traveling when it hits the ground?

Answer:

You can use conservation of energy or the equations of kinematics to solve this. If you choose conservation of energy:

$$mgh = \frac{1}{2}mv^2$$

Solving for v, we get: $v = \sqrt{2gh} = \sqrt{2(9.8)(7.5)} = 12.12$ m/s.

If you choose kinematics:

$$v_f = \sqrt{v_i^2 + 2gy} = \sqrt{0 - 2(9.8)(-7.5)} = 12.12 \text{ m/s.}$$

(b)   With the help of a friend, the boy measures the height to which the ball returns and finds it to be 5.2 meters. How fast was the ball moving when it left the ground?

Answer:

Solving the same way as part a:

$$v = \sqrt{2gh} = \sqrt{2(9.8)(5.2)} = 10.1 \text{ m/s.}$$

(c)   How much mechanical energy was lost in the process?

Answer:

Energy of the ball before hitting the ground:

$$mgh = mg(7.5) = \frac{1}{2}m(12.12)^2 = 29.4 \text{ J}$$

Energy of the ball after hitting the ground:

$$mg(5.2) = \frac{1}{2}m(10.1)^2 = 20.38 \text{ J}$$

The total energy lost equals 29.4 J – 20.38 J = 9.02 J

The percentage lost, then, is $\dfrac{9.02 \text{ J}}{29.4 \text{ J}} \cdot 100\% = 30.7\%$

(d)   Where did the energy go?

Answer:

It was converted into sound (you can hear the ball bounce) and thermal energy of molecules (raising the temperature of the ball and the earth **very** slightly).

IV. LINEAR MOMENTUM

Any moving object has *momentum*, or the product of the *mass* of an object and its *velocity*. To change the momentum of an object, we must apply a force during a time interval. The product of the force acting on an object and the time during which the force acts is called the *impulse*. Thus, the impulse applied to an object changes its momentum. If there are no external forces acting on a system of objects, then the vector sum of their momenta will remain constant; the total momentum will be *conserved*.

Key Terms

center of mass: the point at which all of the mass of an object can be considered concentrated

closed, isolated system: collection of objects in which no forces, matter, or energy can enter or exit the collection

elastic collision: a collision in which both momentum and kinetic energy are conserved

impulse: the product of the force acting on an object and the time over which it acts

impulse-momentum theorem: the impulse imparted to an object is equal to the change in momentum it produces

inelastic collision: a collision in which only momentum is conserved, and not kinetic energy

law of conservation of momentum: the total momentum of a system remains constant during a process

linear momentum: the product of the mass of an object and its velocity

Momentum

The *momentum* of an object is defined as the product of its *mass* and the *velocity* at which it is moving. The usual symbol for momentum is **p**, and thus the equation for momentum is

$$\mathbf{p} = m\mathbf{v}$$

Since mass is a scalar and velocity is a vector, momentum is also a vector, and it is always in the same direction as the velocity. Momentum can be thought of as "inertia in motion," and can be measured by how difficult it is either to move a mass from rest or to stop an object that is already moving. Since the momentum of an object includes both its mass and its velocity, two different masses can have the same momentum depending on their respective velocities.

Impulse

Newton's second law states that an unbalanced (net) force acting on a mass will accelerate the mass in the direction of the force. Another way of saying this is that a net force acting on a mass will cause the mass to change its momentum. We can rearrange the equation for Newton's second law to emphasize the change in momentum:

$$F_{net} = ma = m\frac{\Delta v}{\Delta t}$$

$$F\Delta t = m\Delta v = m(v_f - v_o)$$

The left side of the equation ($F\Delta t$) is called the impulse, and the right side ($m\Delta v$) is the change in momentum. This equation reflects the impulse-momentum theorem, and can be stated: *a force acting on a mass during a time causes the mass to change its momentum.*

Consider the force vs. time graph below:

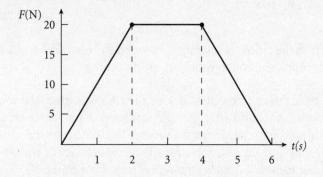

The impulse represented by the graph is the product of the force acting and the time through which it acts. Thus, the impulse is equal to the area under the graph:

Impulse = $F\Delta t$ = *Area under the graph*

Impulse = $\frac{1}{2}(20\ N)(2\ s) + (20\ N)(4\ s - 2\ s) + \frac{1}{2}(20\ N)(6\ s - 4\ s)$

Impulse = 80 N·s.

The impulse-momentum theorem states that the change in momentum of an object is equal to the impulse applied. Thus,

change in momentum = $F\Delta t$ = 80 N·s.

If we know the impulse and change in momentum of an object we can solve for initial or final velocity, or the change in the velocity of the object on which the impulse is applied.

Conservation of Momentum in One Dimension

We've seen that if you want to change the momentum of an object or a system of objects, Newton's second law says that you have to apply an unbalanced force. This implies that if there are no unbalanced forces acting on a system, the total momentum of the system must remain constant. This is another way of stating Newton's first law, the law of inertia. If the total momentum of a system remains constant during a process, such as an explosion or collision, we say that the momentum is conserved. There are three typical examples of conservation of momentum questions on the AP Physics B exam: recoil, inelastic collision, and elastic collision.

Recoil

Consider a rifle of mass m_R which is initially at rest and contains a bullet of mass m_b.

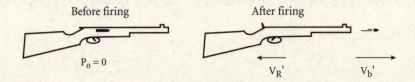

The bullet is fired from the rifle at a velocity v_b' toward the east. The recoil velocity v_R' of the rifle can be found by applying conservation of momentum.

Since the rifle and bullet were motionless before the rifle was fired, the total momentum of the system is zero both before and after the rifle was fired. All forces in this case are internal, and don't change the total momentum of the system. We write the momentum for each mass before and after the rifle is fired, and solve for the recoil velocity of the rifle. We will denote any velocity or momentum that occurs after a process with a prime (′).

$$p_o = p_b' + p_R'$$

$$0 = m_b v_b' + m_R v_R'$$

$$m_b v_b' = -m_R v_R'$$

$$v_R' = -\left(\frac{m_b v_b'}{m_R}\right)$$

Note that the recoil velocity of the rifle is negative, indicating that it moves in the opposite (west) of the bullet, as we would expect. If the rifle is one hundred times more massive than the bullet, the velocity of the rifle is one-hundredth of the velocity of the bullet. In other words, the momentum of the bullet must be equal and opposite to the momentum of the rifle.

Inelastic Collision

An inelastic collision occurs when two colliding objects stick together after impact. Momentum is conserved in an inelastic collision, but kinetic energy is not.

Consider a car of mass m_c moving at a speed v_c that collides with a stationary truck of mass m_T. The two vehicles lock together on impact.

Before the Inelastic Collision

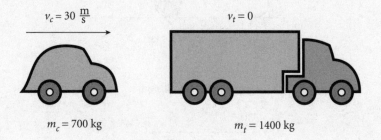

After the Inelastic Collision

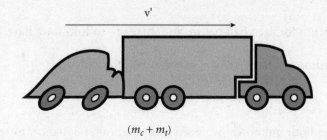

If the combined velocity of the car and truck after the collision is v', we can relate the momentum of the system before the collision to its momentum after the collision:

$$\text{momentum before collision} = \text{momentum after collision}$$
$$m_c v_c = (m_c + m_T) v'$$

We can solve for the velocity v' of the car and truck after the collision.

Elastic Collision

In an elastic collision, the colliding objects bounce off each other, and momentum is conserved. Kinetic energy is also conserved in a perfectly elastic collision.

Consider a white pool ball of mass m_1 moving at a speed v_1 which collides head-on with a red pool ball of mass m_2 initially moving with a speed v_2. After the collision, the white pool ball moves with a speed v_1', and the red pool ball moves with a speed v_2'. Neglecting friction, how would we find the velocity of each ball after the collision?

AP Physics
CHAPTER 5

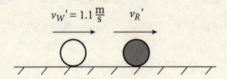

Before the Elastic Collision

$v_W = 10 \frac{m}{s}$ $v_R = 0$

$m_W = 0.5$ kg $m_R = 0.4$ kg

After the Elastic Collision

$v_W' = 1.1 \frac{m}{s}$ v_R'

Writing the momentum for each ball before and after the collision, we have

momentum before collision = momentum after collision

$$m_1 v_1 + m_2 v_2 = m_1 v_1' + m_2 v_2'$$

But if v_1' and v_2' are both unknown, we need to find another equation to solve for them in addition to the conservation of momentum equation. Recall that in a perfectly elastic collision kinetic energy is also conserved. So we can write:

total kinetic energy before collision = total kinetic energy after collision

$$K_1 + K_2 = K_1' + K_2'$$

$$\frac{1}{2}mv_1^2 + \frac{1}{2}mv_2^2 = \frac{1}{2}mv_1'^2 + \frac{1}{2}mv_2'^2$$

We can then solve these equations simultaneously for v_1' and v_2'.

Momentum as a Vector Quantity

Since momentum has both magnitude and direction we must take into account the vector nature of momentum in solving momentum problems in two dimensions. Consider the momentum of the ball in the figure below:

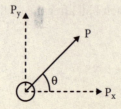

74 KAPLAN

We have resolved the momentum vector **p** into its x- and y- components, p_x and p_y.

$$p_x = p \cos \theta$$

$$p_y = p \sin \theta$$

$$p = \sqrt{(p_x)^2 + (p_y)^2}$$

$$\theta = \tan^{-1}\left[\frac{p_y}{p_x}\right]$$

In a two-dimensional collision, both the x- and y- components of the momentum are conserved. Consider a pool ball striking a second stationary pool ball at an angle, as shown below:

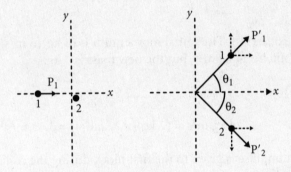

We can write the conservation of momentum equations in each direction:

$$p_x = p_{x'}$$

$$m_1 v_{1x} = m_1 v_{1x}' + m_2 v_{2x}'$$

$$m_1 v_{1x} = m_1 v_1' \cos \theta_1 + m_2 v_2' \cos \theta_2$$

and

$$p_y = p_y'$$

$$0 = m_1 v_{1y}' + m_2 v_{2y}'$$

$$0 = m_1 v_1' \sin \theta_1 + m_2 v_2' \sin \theta_2$$

These equations can be used to solve for velocities before or after the collision, or the angles at which the pool balls move off after the collision.

AP Physics
CHAPTER 5

Sample Problem:

A block slides along a frictionless surface toward a second block attached to the end of a spring, as shown below. When the blocks collide they stick together.

$V = 6$ m/s → $v = 0$ m/s $k = 200$ N/m

$M = 4$ kg $m = 3$ kg

(a) What is the speed of the two blocks immediately after the collision?

Answer:

$$\mathbf{P} = m\mathbf{v}$$

Momentum must be conserved. The initial momentum is (4 kg)(6 m/s), or 24 kg·m/s. After the collision, it must still be 24 kg·m/s, but the new mass is 7 kg.

Thus: $Mv_1 = (M + m)v_2$

$$24 \text{ kg·m/s} = (7 \text{ kg})(v_2), \text{ and } v_2 = 3.43 \text{ m/s}.$$

(b) What impulse is given to the first block during the collision?

Answer:

Impulse is simply change in momentum:

$$\Delta p = m\Delta v = 4 \text{ kg } (3.43 \text{ m/s} - 6.0 \text{ m/s}) = -10.3 \text{ kg·m/s}.$$

[Note: Units for this answer could also be Newton-seconds.]

(c) How far will the spring compress?

Answer:

The energy to compress the spring comes from the kinetic energy of the sliding blocks, which is converted to spring potential energy. We solve for x, the compression of the spring:

$$\frac{1}{2} kx^2 = \frac{1}{2} mv^2$$

$$x = \left(\frac{mv^2}{k} \right) = \frac{(7 \text{ kg})(3.43 \text{ m/s})^2}{(200 \text{ N/m})} = 0.41 \text{ m}$$

(d) What is the maximum speed they will attain once they begin oscillating on the spring?

76 KAPLAN

Physics Review

NEWTONIAN PHYSICS

Answer:

The maximum speed will be attained when all the energy is kinetic. This happens at the point of zero compression, or at the same location where the collision took place.

So the maximum speed will be the same as the speed immediately following the collision: 3.4 m/s.

[Note: You can do this problem by using the Law of Conservation of Energy, as well.]

V. CIRCULAR MOTION AND ROTATION

An object which is moving in a circular path with a constant speed is said to be in *uniform circular motion*. The *angular momentum* of the rotating object is the product of the mass of the object, its speed, and the perpendicular distance from the point of rotation. For an object to move in a circular path, there must be a force exerted on the object which is directed toward the center of the circular path called the *centripetal force*. This centripetal force gives rise to *centripetal acceleration*. A force acting at a perpendicular distance from a rotation point, such as pushing a doorknob and causing the door to rotate on its hinges, produces a *torque*. If the sum of the torques acting on an object is zero, we say that the object is in *equilibrium*, and angular momentum is *conserved*.

Key Terms

angular momentum: the conserved rotational quantity which is equal to the product of the mass, velocity, and radius of motion

centripetal acceleration: the acceleration of an object moving in circular motion which is directed toward the center of the circular path

centripetal force: the central force causing an object to move in a circular path

circular motion: motion of an object moving at a constant radius in a circular path

frequency: the number of vibrations or revolutions per unit of time

hertz: The unit for frequency equal to one cycle or vibration per second

period: the time for one complete cycle or revolution

tangential velocity: the velocity tangent to the path of an object moving in a curved path

torque: the tendency of a force to cause rotation about an axis; the product of the force and the lever arm

uniform circular motion: motion in a circular path of constant radius at a constant speed

KAPLAN 77

Uniform Circular Motion

Up until this chapter we've been reviewing motion in a straight line. The law of inertia states that if an object is moving it will continue moving in a straight line at a constant velocity until a net force causes it to speed up, slow down, or change direction. Imagine a ball moving in a straight line in space, but tied to a string as shown in the figure below.

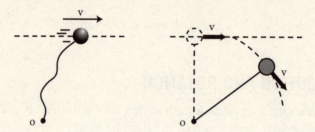

The ball will continue to move in a straight line until it reaches the end of the string, at which time it will be pulled to a central point *O* and begin moving in a circle. As long as the string is pulled toward this central point, the ball will continue moving in a circle at a constant speed. *An object moving in a circle at a constant speed is said to be in uniform circular motion (UCM).* Notice that even though the speed is constant, the velocity vector is not constant, since it is always changing direction due to the central force that the string applies to the ball, which we will call the *centripetal force* $\mathbf{F}_c$. Centripetal means "center-seeking."

There are three vectors associated with uniform circular motion: *velocity* $\mathbf{v}$, *centripetal force* $\mathbf{F}_c$, and *centripetal acceleration* $\mathbf{a}_c$. These vectors are drawn in the diagram below.

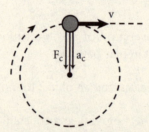

Notice that the velocity vector is tangent to the path of the ball, and points in the direction that the ball would move if the string were to break at that instant. The centripetal force and acceleration are both pointing toward the center. Although the centripetal force does not change the speed of the ball, it is always pulling the ball away from its inertial straight-line path and toward the center of the circle. Thus, the centripetal force accelerates the ball toward the center of the circle, constantly changing its direction.

The time it takes for the ball to complete one revolution is called the *period T*. Since period is a time, we can measure it in seconds, minutes, hours, or even years. On the other hand, *frequency f* is the number of revolutions the ball makes per unit time. Units for frequency would include

$\dfrac{revolutions}{second}$, $\dfrac{revolutions}{hour}$, or any time unit divided into revolutions or cycles. Another name for rev/s is hertz. We can relate all of these quantities in the equations that follow.

The constant speed of any object can be found by using:

$$v = \frac{d}{t}$$

For an object moving in a circle and completing one revolution, the speed of the object can be found by using:

$$v = \frac{circumference}{period} = \frac{2\pi r}{T}, \text{ where } r \text{ is the radius of the circle.}$$

Since frequency is the inverse of the period, we can write:

$$v = 2\pi r f$$

The centripetal acceleration is:

$$a_c = \frac{v^2}{r}$$

Newton's second law states that the net force and acceleration must be in the same direction and they are related by this equation:

$$F_{net} = ma = \frac{mv^2}{r}$$

The centripetal force acting on a ball in UCM pulls the ball inward toward the center. The inertia keeps the ball from moving all the way to the center of the circle. The ball wants to follow a straight line because of its inertia, but the tension in the string pulls it inward away from its straight-line path. *Centripetal* force pulls an object toward the center of its circular path; *centrifugal* force means "center-fleeing", and is not a real force, but the perceived effect of inertia due to the fact that if you are moving in a circular path, you are in a non-inertial reference frame, or an accelerating frame. In a non-inertial reference frame you may be inclined to make up forces to explain the motion inside and outside of your frame of reference. If you feel like you are being pulled outward, you might call what you perceive as an outward force a centrifugal force.

Some other examples of centripetal force are the gravitational force keeping a satellite in orbit, and the friction between a car's tires and the road which causes the car to turn in a circle.

Torque

Torque is the result of a force acting at a distance from a rotational axis, and may cause a rotation about the axis. Torque is the vector product of the displacement vector *r* (as in radius) from the rotational axis and the force **F**:

$$\mathbf{T} = \mathbf{r} \times \mathbf{F}$$

The unit for torque is the *newton-meter*. By the right-hand rule, place your fingers in the direction of **r**, and cross them into the direction of the force **F**, and the torque vector **T** points in the direction of your thumb. To open a hinged door, you apply a force to the doorknob which is mounted a certain distance from the hinges, and create a torque which causes the door to rotate. When you use a wrench to tighten a bolt, you apply a force at the end of the wrench to get the most torque to turn the bolt.

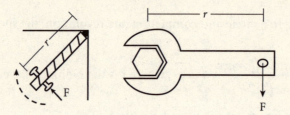

If the force is applied to the wrench so that there is an angle θ between the displacement vector **r** and the force vector **F**, then the magnitude of the torque becomes:

$$T = rF \sin \theta$$

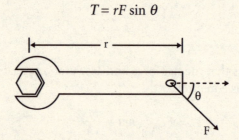

For a torque to be produced there must be a component of the force which is perpendicular to the radius vector. A force and a radius which are parallel to each other will not produce a torque, since the sine of the angle between them is zero.

For a system in static equilibrium, the sum of the forces must equal zero, and the sum of the torques must also equal zero. This is illustrated in the next example.

A 20-kg board, set with its center on a support, serves as a seesaw for two children. One child has a mass of 30 kg and sits 2.5 m from the support. A second child, who has mass 15 kg, sits at the other end of the board at the same distance from the support. Where must a third child, of mass 20 kg, sit in order to balance the board?

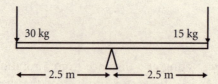

For the board to remain horizontal, the torque on the left must equal the torque on the right. The forces acting on the board on either side are just the weight mg of each child. It should also be noted that since the board is balanced at its center, the weight of the board exerts no torque. It is obvious by looking at the situation that the third child must sit on the same side as the 15 kg child at a distance *x* from the support, to balance the board. So,

$$\text{Torque on the left} = \text{Torque on the right}$$
$$F_1 r_1 = F_2 r_2 + F_3 r_3$$
$$(m_1 g) r_1 = (m_2 g) r_2 + (m_3 g) x$$
$$(30 \text{ kg})(g)(2.5 \text{ m}) = (15 \text{ kg})(g)(2.5 \text{ m}) + (20 \text{ kg})(g)(x)$$

Canceling the g's and solving for *x* we get:

$$x = \frac{(30)(2.5) - (15)(2.5)}{20} = 1.88 \text{ m to the right of the support}$$

Angular Momentum

Linear momentum **p** is the product of the mass of an object and its velocity. For an object moving in a curved path such as a circle, we can also define its *angular momentum*. Consider a ball on the end of a string which is being swung in a circular path. Its angular momentum **L** is defined as the product of its mass, velocity, and radius of orbit:

$$\mathbf{L} = m\mathbf{r} \times \mathbf{v}$$

We are interested in angular momentum because as long as there are no external torques, the quantity *mvr* is conserved. For example, consider a ball on the end of a string which is passed through a glass tube and swung in a circle at a large radius *R*. If we pull the string through the tube to shorten the radius to r, the speed must increase to make up for the loss in radius to conserve angular momentum:

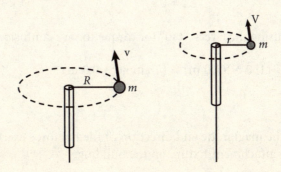

angular momentum before = angular momentum after

$$mvR = mVr$$

We will revisit the law of conservation of angular momentum when we review gravitation and satellite orbits.

CHAPTER 5

Sample Problem:

A horizontal sign pole is to be installed on the front of an office building as shown below. The pole itself weighs 150 N and is 2 meters long. The support wire is attached at a 60° angle at a point 0.5 meters from the building.

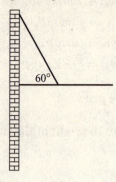

(a) Find the value of the tension in the support cable.

Answer:

Setting the pivot at the point where the sign pole attaches to the building, there are two torques acting. The first is the clockwise torque due to the pole's own weight (150 N), which acts at its exact center, or 1 meter from the pivot. The second is the counterclockwise torque produced by the tension in the support wire, which acts at an angle of 60°, 0.5 meters from the axis of rotation. Set these torques equal, since they must be in rotational equilibrium:

[Note: Forces from the attachment fixture acting at the pivot point do not figure in here, since forces acting through the pivot have a zero lever arm.]

$$\Sigma \tau = 0$$

Note here that we are using the Greek "tau" for torque to save confusion with T for tension.

$$(150 \text{ N})(1.0 \text{ m}) = (T \sin 60°)(0.5 \text{ m})$$

Thus, $T = 346.4$ N.

(b) Find the magnitude and direction of the net force exerted on the pole by the attachment fixture on the building.

Answer:

The flagpole is also in translational equilibrium, so all the forces must cancel in both x and y directions.

First look at the x direction:

$$\Sigma F_x = 0$$

The two forces acting are the component of T that points toward the building and the component of the force of the fixture that cancels it.

$$F_x = T \cos 60° = (346.4)(0.5) = 173.2 \text{ N}$$

Now look at y:

There are three forces acting. The first is the weight of the pole itself (150 N). Then there's the upward force of the tension ($T \sin 60°$). Finally there's the y-component of the force of the fixture: F_y

Add all these up and you get zero: $\Sigma F_y = 0$

$$-150 + T \sin 60° + F_y = 0$$

And $F_y = 150$ N upward (confirmed by the fact that it turned out to be a positive number)

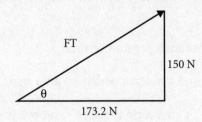

Finally, we combine the two force components F_x and F_y using the Pythagorean Theorem, and get:

$$F_x^2 + F_y^2 = F_t^2 = 173.2^2 + 150^2 = 52{,}498$$

And $F_t = 229.1$ N

The direction is determined by using trig. The relationship between x and y components looks like:

$$\tan \theta = \frac{F_y}{F_x} = \frac{(150 \text{ N})}{(173.2 \text{ N})}$$

$$\theta = \tan^{-1}\left(\frac{150}{173.2}\right) = 40.9°$$

VI. OSCILLATIONS AND GRAVITATION

An object such as a pendulum or a mass on a spring is *oscillating* or vibrating if it is moving in a repeated path at regular time intervals. We call this type of motion *harmonic motion*. For an object to continue oscillating there must be a *restoring force* continually trying to restore it to its equilibrium position.

Key Terms

amplitude: maximum displacement from equilibrium position; the distance from the midpoint of a wave to its crest or trough

ellipse: an oval-shaped curve which is the path taken by a point that moves such that the sum of its distances from two fixed points (foci) is constant; the planets move in elliptical orbits around the Sun

equilibrium position: the position about which an object in harmonic motion oscillates; the center of vibration

frequency: the number of vibrations per unit of time

gravitational field: space around a mass in which another mass will experience a force

gravitational force: the force of attraction between two objects due to their masses

inverse square law: situation where one physical quantity varies as the inverse square of the distance from its source

law of universal gravitation: the gravitational force between two masses is proportional to the product of the masses and inversely proportional to the square of the distance between them.

mechanical resonance: condition in which natural oscillation frequency equals frequency of a driving force

period: the time for one complete cycle of oscillation

periodic motion: motion that repeats itself at regular intervals of time

restoring force: the force acting on an oscillating object which is proportional to the displacement and always points toward the equilibrium position.

satellite: any object which orbits another more massive object, such as the moon orbiting the earth

The Dynamics of Harmonic Motion

Another type of motion commonly found in nature is oscillatory, or vibrational, motion, which we sometimes call harmonic motion. An object is in harmonic motion if it follows a repeated path at regular time intervals. Two common examples of harmonic motion often studied in physics are a mass on a spring and a pendulum.

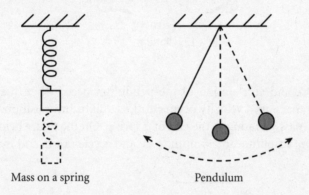

Mass on a spring   Pendulum

As an object vibrates, it has both a *period* and a *frequency*. Recall that the period of vibration is the time it takes for one complete cycle of motion, that is the time it takes for the object to return to its original position. For a pendulum, this would be the time for one complete swing, up and back. The frequency is the number of cycles per unit time, such as cycles per second, or hertz. The period and frequency of a pendulum depend only on the length of the pendulum and the acceleration due to gravity. The lowest point in the swing of a pendulum is called the *equilibrium position*, and the maximum displacement from equilibrium is called the *amplitude*. The amplitude can be measured by the maximum angle or by the linear horizontal distance from equilibrium, as shown below.

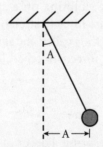

Since the pendulum vibrates about the equilibrium position, there must be a force that is trying to restore it back toward the center of the swing. This force is called *the restoring force*, and it is greatest at the amplitude and zero as the pendulum passes through the equilibrium position. Newton's second law tells us that if there is a net force, there must be an acceleration, and if the force is maximum at the amplitude, the acceleration must be maximum at the amplitude as well. The velocity, however, is zero at the amplitude and maximum as it passes through the equilibrium position.

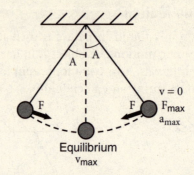

The maximum force and acceleration of the pendulum occur when the pendulum is at its amplitude, and the maximum velocity occurs at the equilibrium position. We can apply these same concepts to a mass vibrating at the end of a spring. On the figure below, we've labeled the equilibrium position, amplitude, maximum force and acceleration, and maximum velocity.

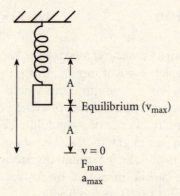

The period and frequency of a mass vibrating on a spring depend on the stiffness in the spring. For a stiffer spring, it takes more force to stretch the spring to a particular length. The amount of force needed per unit length is called the *spring constant k*, measured in newtons per meter. The relationship between force, stretched length, and k for an ideal (or linear) spring is called Hooke's Law:

$$F_{spring} = -kx$$

where x is the stretched length of the spring. For an ideal spring, the stretch is proportional to the force, but in the opposite direction. If we pull with twice the force, the spring will stretch twice as far.

The graph below represents the magnitude of the force *F vs. stretch length x* for two springs:

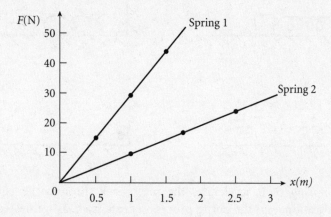

The slope of each line is the change in force (rise) divided by the change in length (run). Since this ratio is also equal to the spring constant k, the graph with the higher slope has the higher spring constant, which is an indication of the stiffness of the spring. Spring 1 is the stiffer spring, since it takes more force to stretch it the same length as Spring 2.

We can find the spring constant k for Springs 1 and 2 by taking the ratio of the force to the stretch for a particular interval. In other words, we can find the slope of the *F* vs. *x* graph for each spring. The slope of the line and the spring constant for Spring 1 is 30 N/m, and the spring constant for Spring 2 is 10 N/m.

Energy Considerations in Harmonic Motion

As an object vibrates in harmonic motion, energy is transferred between potential energy and kinetic energy. Consider a mass sitting on a surface of negligible friction and attached to a linear spring. If we stretch a spring from its equilibrium (unstretched) position to a certain displacement, we do work on the mass against the spring force. By the work-energy theorem, the work done is equal to the stored potential energy in the spring. If we release the mass and allow it to begin moving back toward the equilibrium position, the potential energy begins changing into kinetic energy. As the mass passes through the equilibrium position, all of the potential energy has been converted into kinetic energy, and the speed of the mass is maximum. The kinetic energy in turn begins changing into potential energy, until all of the kinetic energy is converted into potential energy at maximum compression.

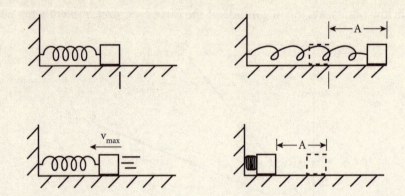

The compressed spring then accelerates the mass back through the equilibrium to the original starting position, and the entire process repeats itself. If we neglect friction on the surface and in the spring, the total energy of the system remains constant, that is:

Total Energy = Potential Energy + Kinetic Energy = constant

Thus, whatever potential energy is lost must be gained by kinetic energy, and vice-versa. As long as no energy is lost to the surroundings, the mass on the spring continues to oscillate. The same would be true for a pendulum.

As the spring oscillates, we can calculate the total mechanical energy at any time:

$$E_{total} = K + U$$

where $K = \frac{1}{2}mv^2$ and $U = \frac{1}{2}kx^2$

In the equation for the potential energy of the spring above, x is the displacement from equilibrium position at any time. Because of the oscillatory nature of the vibrating mass, we can express the displacement x from equilibrium position at any time t as

$$x = A\cos(\omega t)$$

where A is the amplitude of oscillation and ω (the lower-case Greek letter omega) is called angular frequency, and is measured in radians per second. The mass on the spring makes one full oscillation (2π radians) in one period T, so the angular frequency can be found by

$$\omega = \frac{2\pi}{T} = 2\pi f, \text{ where } f \text{ is the frequency in hertz.}$$

Another relationship between the angular frequency of a mass oscillating on a spring and the spring constant k is:

$$\omega = \sqrt{\frac{k}{m}}$$

We can see from this equation that the higher the spring constant k, the stiffer the spring, and the greater the angular frequency of oscillation. A smaller mass will also increase the angular frequency for a particular spring.

If we set the two equations above for ω equal to each other and solve for the period T of oscillation, we get:

$$T = 2\pi\sqrt{\frac{m}{k}}$$

Similarly, the equation for the period of a pendulum is

$$T = 2\pi\sqrt{\frac{l}{g}}$$

where l is the length of the pendulum, and g is the acceleration due to gravity at the location of the pendulum. A longer length will have a longer period, while a stronger gravitational field will shorten the period of a pendulum.

Gravitation

Newton's law of universal gravitation states that all masses attract each other with a gravitational force which is proportional to the product of the masses and inversely proportional to the square of the distance between them. The gravitational force holds *satellites* in orbit around a planet or star.

Gravitational Force

Newton's law of universal gravitation states that every mass in the universe attracts every other mass in the universe. The gravitational force between two masses is proportional to the product of the masses and inversely proportional to the square of the distance between their centers. The equation describing the gravitational force is

$$F_G = \frac{Gm_1m_2}{r^2}$$

where F_G is the gravitational force, m_1 and m_2 are the masses in kilograms, and r is the distance between their centers. The negative sign indicates that it is an attractive force. The constant G simply links the units for gravitational force to the other quantities, and in the metric system happens to be equal to 6.67×10^{-11} Nm^2/kg^2. Like several other laws in physics, Newton's law of universal gravitation is an inverse square law, where the force decreases with the square of the distance from the centers of the masses. For example, if the force between the two masses shown in the figure above is F, then if we double the distance between the masses, the force will be only one-quarter as much. If we triple the distance between them, the force will be one-ninth as much, and so on.

Gravitational Acceleration

We've been treating the acceleration due to gravity as if it were a constant as long as we stay near the surface of the Earth. But if we travel an appreciable distance from the center of the Earth, such as twice the Earth's radius, the acceleration due to gravity is changed significantly. By Newton's second law, gravitational acceleration is proportional to gravitational force. Like the gravitational force, the gravitational acceleration is inversely proportional to the square of the distance between the centers of the two masses.

For example, we know that a ball dropped near the surface of the Earth accelerates toward the Earth with an acceleration of 9.8 m/s². With what initial acceleration will the ball fall if it is dropped from a height of three Earth radii from the center of the Earth?

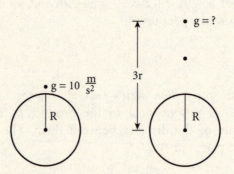

Since the acceleration due to gravity is inversely proportional to the square of the distance from the center of the Earth, we can write:

$g \propto \dfrac{1}{r^2}$, where the $\propto$ sign means "proportional to." If the ball is dropped at a height of three earth radii, we have:

$$g \propto \dfrac{1}{(3r)^2} = \dfrac{1}{9r^2}$$

Thus, the acceleration due to gravity at three earth radii is $\dfrac{1}{9}$ as much as it is near the earth's surface (one earth radius). The value for g is then $\dfrac{1}{9}$ (9.8 m/s²) = 1.1 m/s² at this height.

We can derive the equation for calculating the gravitational acceleration at any height above any planet if the mass of the planet is known. The weight of any object at any point in a gravitational field is equal to the gravitational force acting on the mass *m* of the object:

$$\text{Weight} = F_G$$

Setting the equations for weight and gravitational force equal to each other gives us

$$mg = \frac{-GmM}{r^2}$$

where m is the mass of the object in the gravitational field, and M is the mass of the source of the gravitational field, such as a planet or star. After canceling the mass m of the object, we see that the gravitational acceleration at a distance r from the center of a planet of mass M is

$$g = \frac{-GM}{r^2}$$

The convention used here, with the negative, indicates that g is in the same direction as F_G—toward the center of the source (such as Earth).

Orbiting Satellites

As a satellite such as the moon orbits the earth, it is pulled toward the earth with a gravitational force which is acting as a centripetal force. As described in an earlier section, the inertia of the satellite causes it to tend to follow a straight-line path, but the centripetal gravitational force pulls it toward the center of the orbit.

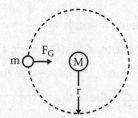

If a satellite of mass m moves in a circular orbit around a planet of mass M, we can set the centripetal force equal to the gravitational force (both toward the center of the planet) and solve for the speed of the satellite orbiting at a particular distance r:

$$F_c = F_G$$

$$\frac{mv^2}{r} = \frac{GmM}{r^2}$$

$$v = \sqrt{\frac{GM}{r}}$$

Angular Momentum and Energy of an Orbiting Satellite

Although we often approximate the orbits of satellites around the earth (or around the sun, in the case of the orbits of the planets) as circular, they are actually elliptical, with the earth (or sun) located at one focus of the ellipse.

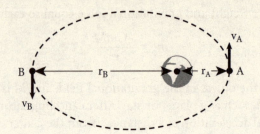

This means that there are times when the satellite is closer to the earth than other times. If the satellite is at point A in the figure above, the gravitational force acting on the satellite is greater than at point B, and the speed at A is greater than at B. This is consistent with the law of conservation of angular momentum.

Angular Momentum at A = Angular Momentum at B

$$mv_A r_A = mv_B r_B$$

This equation shows that the closer the satellite gets to the planet (smaller r), the faster it goes (larger v). As the satellite orbits, potential and kinetic energies each change, but total energy (potential plus kinetic) remains constant.

The kinetic energy of an orbiting satellite is simply $K = \frac{1}{2}mv^2$ at any point along the orbit. But the potential energy of the satellite cannot be found by $U = mgh$. This equation is only valid at distances very close to the surface of a planet. When two masses, like the earth and the moon, are separated by an appreciable distance r compared to the radius of the masses, we must use the equation

$$U_G = -\frac{Gm_1 m_2}{r}$$

to find the gravitational potential energy between the two masses m_1 and m_2. In this equation, the potential energy U between the two masses is defined to be zero when the masses are an infinite distance apart, since they apply no force to each other at this distance. Thus, the potential energy would be negative for any distance r which is smaller than infinity. The total energy of a satellite of mass m orbiting a large planet of mass M at a distance r is

$$E_{total} = K + U_G = \frac{1}{2}mv^2 + \frac{-GmM}{r}$$

Consider a satellite orbiting the earth moving from point A to point B in an elliptical orbit as shown below:

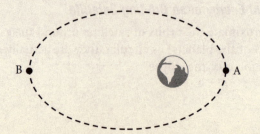

NEWTONIAN PHYSICS

Fill in the table below with labels "increases," "decreases," or "constant" to describe the quantities related to the orbit, as the satellite moves in its orbit to a position farther from earth.

Quantity	Increases, Decreases, Constant
Speed	
Angular Momentum	
Kinetic Energy	
Potential Energy	
Total Energy	
Gravitational Force	

Answers:

Quantity	Increases, Decreases, Constant	Explanation
Speed	Decreases	The satellite is moving farther away from the planet and is slowing down.
Angular momentum	Constant	Conservation of angular momentum.
Kinetic energy	Decreases	Since the speed is decreasing, the kinetic energy must also be decreasing.
Potential energy	Increases	Since the potential energy is negative, a larger orbital distance r would give a smaller negative value for U. Also, if kinetic energy is decreasing, potential energy must be increasing, since total energy must remain constant.
Total energy	Constant	Conservation of total energy.
Gravitational Force	Decreases	The satellite is getting farther away, thus the force is getting weaker by the inverse square law.

AP Physics

CHAPTER 5

Sample Problem:

A satellite is placed in orbit 100 km above the surface of the earth (radius = 6,370 km, mass = 5.98×10^{24} kg).

(a) What is the value of the acceleration of gravity at that altitude as a percentage of the acceleration at the surface of the earth?

Answer:

The acceleration of gravity $= \dfrac{F_g}{m_s} = \dfrac{Gm_E}{r^2} = \dfrac{(6.67 \times 10^{-11})(5.98 \times 10^{24})}{(6.37 \times 10^6 + 100,000)^2} = 9.53$ m/s^2

Thus, the acceleration of gravity there is $\dfrac{9.53}{9.8} = 0.97$ g

(b) If the satellite has a mass of 1,500 kg, what is the force of gravity acting on it?

Answer:

$$F = ma = (1{,}500 \text{ kg})(9.53 \text{ m/s}^2) = 14{,}295 \text{ N}$$

Alternate solution: $F_g = \dfrac{Gm_E m_s}{r^2} = \dfrac{(6.67 \times 10^{-11})(5.98 \times 10^{24})(1{,}500)}{(6.43 \times 10^6)^2} = 14{,}295$ N

(c) What is the orbital period of the satellite?

Answer:

At that altitude, we know the centripetal acceleration (acceleration of gravity) is 9.53 m/s^2, from part (a). Since $a_c = \dfrac{v^2}{r}$, we can find the orbital speed as:

$$v = \sqrt{a_c r} = \sqrt{(9.53 \text{ m/s}^2)(6.43 \times 10^6 \text{ m})} = 7{,}828 \text{ m/s}$$

The length of one orbit is the circumference of the circle it makes, and therefore:

$$T = \frac{\text{distance}}{\text{speed}} = \frac{(2\pi)(6.43 \times 10^6 \text{ m})}{7{,}828 \text{ m/s}} = 5{,}161 \text{ s} = 1.43 \text{ hours}$$

(d) If NASA wanted to drop its orbital altitude to 60 km, would it have to speed up or slow down? Explain your response.

Answer:

If the altitude drops to 60 km, the gravitational force will increase to 14,470 N. Since this force produces centripetal acceleration, we use it to find the new speed.

$F = \dfrac{mv^2}{r}$ leads us to the fact that $v = \sqrt{\dfrac{Fr}{m}} = \sqrt{\dfrac{14{,}470 \text{ N} \cdot 6.43 \times 10^6 \text{ m}}{1{,}500 \text{ kg}}} = 7{,}875$ m/s.

Therefore the speed must increase in order to orbit at a lower altitude.

NEWTONIAN PHYSICS

Important Equations

Definition of weight (force of gravity on an object): $\quad W = mg$

Definition of static friction: $\quad F_s \leq \mu_s N$

Definition of kinetic friction: $\quad F_k = \mu_k N$

Potential energy of a spring: $\quad U_s = \left(\dfrac{1}{2}\right)kx^2$

Work: $\quad W = \boldsymbol{F} \cdot \boldsymbol{s} = Fs \cos \theta$

Hooke's Law: $\quad \boldsymbol{F}_s = -k\boldsymbol{x}$

Kinetic energy: $\quad K = \left(\dfrac{1}{2}\right)mv^2$

Gravitational potential energy: $\quad U_g = mgh$

Work-Energy Theorem: $\quad W = \Delta KE = KE_f - KE_i$

Efficiency: $\quad \text{Eff} = \dfrac{\text{Work out}}{\text{Energy in}}$

Momentum: $\quad \mathbf{P} = m\boldsymbol{v}$

Impulse-Momentum Theorem: $\quad \mathbf{J} = \Delta \boldsymbol{p} = \boldsymbol{F}\Delta t = m\Delta \boldsymbol{v}$

Displacement: $\quad \boldsymbol{s} = \Delta x = x - x_0$

Distance: $\quad d = |\Delta x| = |x - x_0|$

Average velocity: $\quad \boldsymbol{v} = \dfrac{\boldsymbol{s}}{t}$

Average acceleration: $\quad \boldsymbol{a} = \dfrac{\Delta v}{\Delta t}$

Instantaneous velocity $\quad v = \lim\limits_{t \to 0} \dfrac{\Delta x}{\Delta t}$

Instantaneous acceleration $\quad a = \lim\limits_{t \to 0} \dfrac{\Delta v}{\Delta t}$

Power: $\quad P_{\text{ave}} = \dfrac{W}{\Delta t} = \mathbf{F} \cdot \mathbf{v} = Fv \cos \theta$

Average velocity: $\quad v_{\text{av}} = \left(\dfrac{1}{2}\right)(v + v_0)$

KAPLAN 95

AP Physics

CHAPTER 5

Accelerated motion:	$v_f = v_i + at$
	$v_f^2 = v_i^2 + 2a(x_f - x_i)$
	$x_f = x_i + v_o t + \dfrac{1}{2}at^2$
Arc length:	$s = r\theta$
Angular velocity:	$\omega = \dfrac{\Delta\theta}{\Delta t}$
Tangential velocity:	$v = r\omega$
Centripetal acceleration:	$a_c = \dfrac{v^2}{r}$
Centripetal force:	$F_c = \dfrac{mv^2}{r}$
Newton's Universal Law of Gravitation:	$F_g = -\dfrac{(Gm_1 m_2)}{r^2}$
Gravitational potential energy:	$U_g = -\dfrac{(Gm_1 m_2)}{r}$
Newton's Second Law (linear):	$\Sigma \mathbf{F} = \mathbf{F}_{net} = \mathbf{ma}$
Torque:	$\tau = \mathbf{r} \times \mathbf{F} = rF\sin\theta$

PRACTICE PROBLEMS

1. A car travels around a flat, circular turn of radius 50 meters while maintaining a constant speed of 15 m/s.

    (a) Draw and clearly label a force diagram showing all forces acting on the car in this problem.
    (b) Find the minimum value for the coefficient of friction necessary to keep the car on the road.
    (c) In icy conditions, the coefficient of friction drops to 0.15. Now find the maximum safe speed for making the turn.
    (d) If the flat roadway were replaced by a banked one, explain briefly the difference in your calculation of maximum safe speed.

2. Consider the pulley apparatus below.

    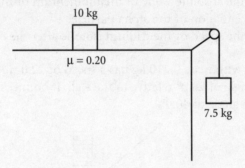

    (a) Find the acceleration of the two-block system.
    (b) What is the tension in the rope?
    (c) Now suppose the hanging mass is *replaced* by a downward force of 75 N. What will be the resulting acceleration of the system?
    (d) What is the new tension?
    (e) How much mass should be added to the first block for it to slide at constant velocity?

3. A 25,000 kg train car collides with an identical train car traveling in the same direction. The first car's initial speed was 1.2 m/s, and the second's was 0.4 m/s.

    (a) If they stick together upon collision, what is their speed?
    (b) Is mechanical energy conserved? Justify your answer.
    (c) How much force will it take to stop the moving cars in a distance of 100 meters?

AP Physics
CHAPTER 5

4. A skier goes down a slope with an angle of 35 degrees relative to the horizontal. Her mass, including all equipment, is 70 kg. The coefficient of kinetic friction between her skis and the snow is 0.15.

    (a) Draw a free-body diagram of the skier.
    (b) Calculate the net force acting on the skier.
    (c) If the slope is 60 m long, what is her speed at the bottom of the slope, assuming that she started from rest?

5. A gold atom has a mass of 3.287×10^{-25} kg and is at rest. It emits an alpha particle that has a mass of 6.646×10^{-27} kg. In this process, the gold atom is transformed into an iridium isotope. The alpha particle has a speed of 3.40×10^6 m/s.

    (a) What is the absolute value of the momentum of the alpha particle?
    (b) What is the absolute value of the momentum of the iridium isotope after the emission of the alpha particle?
    (c) What is the speed of the iridium isotope after the emission process?

6. A billiard ball with mass 0.210 kg has a speed of 22.0 m/s and collides with a wall at an angle of 28.0° relative to the wall. It bounces off with a speed of 17.0 m/s at an angle of 40.0°.

    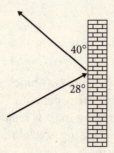

    (a) What is the impulse that the wall imparts to the ball?
    (b) If the ball is in contact with the wall for 0.001 second, what force does the ball exert on the wall?
    (c) What is the loss in mechanical energy due to the ball's collision with the wall?

7. A fire hose sprays 450 liters of water a minute onto a fire with a velocity of 17.0 m/s. Water has a density of 1 kg/liter.

    (a) If the hose is aimed upward at a 45° angle toward a building, how high does the water reach?
    (b) What is the average force, due to the water, acting on the fireman holding the hose?

8. A crane is used to hold a mass of 230 kg as shown in the picture. The crane consists of a horizontal steel beam that is supported by the vertical tower and held in place by a steel cable (on the right side).

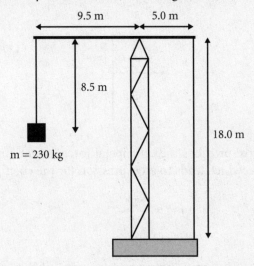

   (a) Draw a diagram showing all forces acting on the horizontal steel beam.
   (b) Calculate the tension in the steel cable on the right.
   (c) Calculate the force that the vertical tower exerts on the beam.

9. Imagine it's the year 2020 and you are one of the select few who get to live on a space colony. The colony vehicle is spherical in shape, with a mass of about 8.896×10^6 kg and a radius of 6,120 meters. It rotates about its vertical axis so that the speed on the surface of the sphere is 300 mph, or about 134.1 m/s.

   (a) What gravitational acceleration, in terms of g, is felt at the surface of the sphere?
   (b) If the colony is in a circular orbit around the earth 100 km away, how fast is it moving in its orbit? The radius of the earth is approximately 6,387.5 km at the equator.
   (c) How long does it take the colony to orbit the Earth?

10. A spring-loaded dart gun (k = 350 N/m) is fired, and a 15 g dart leaves the barrel traveling at 12.5 m/s.

    (a) How far was the spring compressed before the gun was fired?
    (b) The dart flies through the air and sticks to a metal block that is free to slide along a frictionless table. The block has a mass of 1 kg. Find the final speed of the dart and block.
    (c) How much energy is lost in the collision?

AP Physics
CHAPTER 5

Solutions to Practice Problems

1. Answers:

(a)

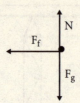

(b) Since the friction force provides the centripetal force for the turn, we set friction force equal to centripetal force, which leads to an expression for the coefficient of friction:

$$\mu mg = \frac{mv^2}{r}$$

$$\mu = \frac{v^2}{gr} = \frac{(15 \text{ m/s})^2}{(9.8 \text{ m/s}^2)(50 \text{ m})} = 0.46$$

(c) Again, using the relationship shown above between speed and coefficient of friction: $v = \sqrt{\mu gr} = \sqrt{(0.15)(9.8)(50)} = 8.6$ m/s

(d) When the road is banked, the centripetal force is provided by both the friction force and a normal force component that is now directed toward the center of the arc of the car's turn. The car is able to increase speed somewhat in the turn with the banked road.

2. Answers:

(a) The net force on the system equals the downward force of gravity on the 7.5 kg block (73.5 N) minus friction ($F_f = \mu mg = 0.2(10)(9.8) = 19.6$ N).

Thus $F_{net} = 73.5 - 19.6 = 53.9$ N

Then acceleration equals $\frac{F}{m}$:

$$a = \frac{F}{m} = \frac{53.9 \text{ N}}{17.5 \text{ kg}} = 3.08 \text{ m/s}^2$$

Note that with this method we must use the total mass of the system.

(b) The tension minus friction has to equal the net force on the 10 kg block. We know friction equals 19.6 N and given the acceleration of 3.08 m/s², we know that the net force must be 30.8 N ($F = ma$). Thus we have:

$T - 19.6$ N $= 30.8$ N which means that $T = 50.4$ N

100 KAPLAN

(c) The same process as part *a* will hold except that the downward force will now equal 75 N and the total mass of the system now equals 10 kg (no second mass to accelerate).

Friction remains constant, so the net force = 75 N – 19.6 N = 55.4 N.

Acceleration, then, equals $\dfrac{F}{m} = \dfrac{55.4 \text{ N}}{10 \text{ kg}} = 5.54 \text{ m/s}^2$.

(d) Now, the only force acting on the block besides friction is the tension, and it comes from the 75 N downward force. Thus, the new tension is 75 N. Combined with the friction force of 19.6 N in the other direction, the net force of 55.4 N is easy to find.

(e) For the block not to accelerate, the net force on it must be zero. There are two forces acting on it in the horizontal direction: friction and tension. The 75 N tension force won't change, so we need to find the mass required for friction to equal 75 N.

$$F_f = 75 \text{ N} = \mu\, mg = 0.2(x)(9.8)$$

Solving this for our unknown mass x, we get a total mass of 38.3 kg. Since we already have 10 kg, we will need an additional 28.3 kg.

3. Answers:

(a) Momentum must be conserved before and after the collision:

$$p = mv$$
$$mv_1 + mv_2 = 2mV$$
$$(25{,}000 \text{ kg})(1.2 \text{ m/s}) + (25{,}000 \text{ kg})(0.4 \text{ m/s}) = (50{,}000 \text{ kg})V$$
$$V = 0.8 \text{ m/s}$$

(b) No. The combined initial kinetic energy of the two cars equals:

$$KE_t = KE_1 + KE_2 = \frac{1}{2}(25{,}000 \text{ kg})(1.2\,\frac{\text{m}}{\text{s}})^2 + \frac{1}{2}(25{,}000 \text{ kg})(0.4\,\frac{\text{m}}{\text{s}})^2 = 18{,}000 + 2{,}000 =$$
20,000 Joules

After the collision, they move as one mass so their total energy equals:

$$KE_t = \frac{1}{2}(50{,}000 \text{ kg})\left(0.8\,\frac{\text{m}}{\text{s}}\right)^2 = 16{,}000 \text{ Joules}$$

Since the initial and final energies aren't equal, total energy isn't conserved.

(c) This part can be worked with either energy or kinematic methods. However, since we already know the kinetic energy from the previous part, it will be easier to use the energy method here. The amount of work it will take to stop the cars is equivalent to the kinetic energy they currently possess, by the Work-Energy Theorem:

$$W = \Delta KE \text{ and } W = Fs$$

$$F = \frac{KE}{s} = \frac{(16{,}000 \text{ J})}{(100 \text{ m})} = 160 \text{ N}$$

4. Answers:

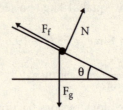

(b) There are two forces acting on her along the direction of the slope. One is the gravity component along the plane, acting downward, and the other is the force of kinetic friction, acting upwards along the slope.

The force of kinetic friction is the friction coefficient times the normal force. The normal force, in turn, is just big enough to balance the component of gravity perpendicular to the slope, so there is no net force in that direction:

N = mg cos θ, where m = 70 kg.

This gives the net force, which is along the slope:

$$F_{net} = mg \sin \theta - \mu_k mg \cos \theta$$
$$= mg(\sin \theta - \mu_k \cos \theta)$$
$$= (70 \text{ kg})(9.81 \text{ m/s}^2)(\sin 35° - 0.15 \cos 35°)$$
$$= 309 \text{ N}$$

(c) Her acceleration is simply:

$$a = g(\sin \theta - \mu_k \cos \theta) = 4.5 \text{ m/s}^2$$

Therefore, we find her speed at the end of the slope:

$$v^2 = v_0^2 + 2a \Delta x = 2(4.5 \text{ m/s}^2)(60 \text{ m}) = 540 \text{ m}^2/\text{s}^2$$

And therefore v = 23.2 m/s.

5. Answers:

(a) $p_{alpha} = m_{alpha} v_{alpha} = (6.646 \times 10^{-27} \text{ kg})(3.40 \times 10^6 \text{ m/s}) = 2.26 \times 10^{-20}$ kg·m/s

(b) No calculation is necessary. The gold nucleus initially had no momentum, because it was at rest. So the whole system of the two decay products, the alpha particle and the iridium isotope, still has 0 momentum. Therefore, the momenta of the two decay products must be equal and opposite. The absolute value of the momentum of the iridium isotope is also 2.26×10^{-20} kg·m/s.

Physics Review

NEWTONIAN PHYSICS

(c) For the purposes of this problem, assume that the mass of the iridium isotope is simply the difference of the mass of the gold atom and that of the alpha particle.

Since we already have the momentum, and we also know the mass, we can calculate the velocity:

$$v = \frac{p}{m} = \frac{(2.26 \times 10^{-20}\ \text{kg·m/s})}{(3.287 \times 10^{-25}\ \text{kg} - 6.646 \times 10^{-27}\ \text{kg})} = 7.02 \times 10^4\ \text{m/s} = 70.2\ \text{km/s}$$

6. Answers:

(a) We need to calculate the change of the velocity components in x- and y-directions separately, remembering that the x-velocity after the collision is in the negative direction.

$$\Delta v_x = v_{xf} - v_{xi} = (-17.0\ \text{m/s})(\sin 40°) - (22.0\ \text{m/s})(\sin 28°) = -21.3\ \text{m/s}$$

$$\Delta v_y = v_{yf} - v_{yi} = (17.0\ \text{m/s})(\cos 40°) - (22.0\ \text{m/s})(\cos 28°) = -6.4\ \text{m/s}$$

[Note: The two y components are in the same direction.]

$$\Delta v = (\Delta v_x{}^2 + \Delta v_y{}^2)^{1/2} = 22.2\ \text{m/s}$$

The impulse is then simply:

$$\Delta p = m\Delta v = (0.21\ \text{kg})(22.2\ \text{m/s}) = 4.66\ \text{kg·m/s}$$

(b) We can calculate the force of the wall on the ball by the formula:

$$F = \frac{\Delta p}{\Delta t} = \frac{(4.66\ \text{kg·m/s})}{(0.001\ \text{s})} = 4,660\ \text{N}$$

By Newton's third law, a force of the same magnitude is exerted by the ball on the wall, but in the opposite direction.

(c) The change in kinetic energy before and after the collision can be calculated:

$$\Delta KE = KE_f - KE_I = \frac{1}{2}mv_f^2 - \frac{1}{2}mv_i^2 =$$

$$= \left(\frac{1}{2}\right)(0.21\ \text{kg})(17\ \text{m/s})^2 - \left(\frac{1}{2}\right)(0.21\ \text{kg})(22\ \text{m/s})^2 = -20.5\ \text{J}$$

This is a loss of 20.5 Joules of kinetic energy during the collision.

7. Answers:

(a) Using the vertical component of the velocity of the water as it comes out of the hose and the knowledge that at the peak of its motion the water's vertical velocity is zero: $v_f^2 = v_i^2 + 2as$

$$0 = (17 \cos 45°\ \text{m/s})^2 + (2)(-9.8\ \text{m/s}^2)(s)$$

KAPLAN 103

The displacement of the water vertically is: s = 7.4 meters

(b) By Newton's third law, the force pushing the water out of the hose has a reaction force on the fireman. The average force is:

$$F_{ave} = \frac{\Delta p}{\Delta t} = \frac{m\Delta v}{\Delta t}$$

At the rate of 450 liters per minute, during a time $\Delta t = 60$ seconds, 450 kg water is emitted. This gives:

$$F_{ave} = \frac{(450 \text{ kg})(17 \text{ m/s})}{(60 \text{ s})} = 127.5 \text{ N}$$

This is a fairly sizeable force and is the reason why firefighters have to rotate frequently in and out of holding the hose

8. Answers:

(a)

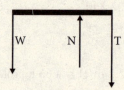

(b) To calculate the tension T, use the fact that the beam is in rotational equilibrium. Set the pivot at the point where the support is attached so that the normal force N from the support exerts no torque. Thus, there are only two forces exerting torques—the weight of the suspended mass and the tension in the cable.

$$\Sigma \tau = 0$$

Clockwise torque: $\tau_{cw} = T (5.0 \text{ m})$

Counterclockwise torque: $\tau_{ccw} = (230 \text{ kg})(9.8 \text{ m/s}^2)(9.5 \text{ m})$

Setting clockwise and counterclockwise torques equal to each other and solving for T:
T = 4,283 N

(c) To find the force of the support, N, on the beam, set up translational equilibrium in the y direction:

$$\Sigma F_y = 0$$
$$N - mg - T = 0$$
$$N = mg + T = (230 \text{ kg})(9.8 \text{ m/s}^2) + 4{,}283 \text{ N} = 6{,}540 \text{ N}$$

Physics Review

NEWTONIAN PHYSICS

9. Answers:

(a) First, knowing the radius and the tangential speed of rotation, we can find the centripetal acceleration (which simulates gravity) at the surface of the station:

$$a_c = \frac{v^2}{r} = 2.94 \text{ m/s}^2.$$

Dividing this by the normal acceleration of gravity (9.8 m/s^2), we find that the answer is 0.3 g's.

(b) Since the only force keeping the space colony vehicle in circular motion is gravity, we set the universal gravitation equation equal to the equation for centripetal force and solve for the speed, remembering that orbital radius is the radius of Earth plus satellite distance.

$$\frac{GM_E m_c}{r^2} = \frac{m_c v^2}{r}$$

$$v = \sqrt{\frac{GM_E}{(r_E + h)}} = 7{,}841 \text{ m/s}$$

(c) The trip is (basically) circular, which makes it a distance of $2\pi r$, or:

$$2(\pi)(6.39 \times 10^6 \text{ m} + 1 \times 10^5 \text{ m})$$

Using $v = \dfrac{2\pi r}{T}$, $T = \dfrac{2\pi r}{v} = \dfrac{4.078 \times 10^7 \text{ m}}{7{,}841 \text{ m/s}} = 5{,}200$ seconds (about 1.5 hours)

10. Answers:

(a) The kinetic energy of the dart is a conversion from the potential energy of the compressed spring:

$$\frac{1}{2} kx^2 = \frac{1}{2} mv^2$$

$$x = \left(\frac{mv^2}{k} \right)^{1/2} = 0.082\text{m or } 8.2 \text{ cm}$$

(b) Momentum must be conserved in the collision, so the momentum of the dart before it hits its target must be equal to the momentum of the combination after the dart hits the object:

$$mv = (m + M)V$$

$$(0.015 \text{ kg})(12.5 \text{ m/s}) = (1.015 \text{ kg})V$$

$$V = 0.18 \text{ m/s}$$

KAPLAN 105

AP Physics

CHAPTER 5

(c) The change in kinetic energy during the collision:

$$\Delta KE = KE_f - KE_I$$

$$= \frac{1}{2}(1.015 \text{ kg})(0.18 \text{ m/s})^2 - \frac{1}{2}(0.015 \text{ kg})(12.5 \text{ m/s})^2$$

$$= -1.16 \text{ J}$$

Thus, 1.16 J of kinetic energy is lost during the collision.

Fluid Mechanics and Thermal Physics

CHAPTER 6

I. FLUID MECHANICS

A *fluid* is any substance that flows, typically a liquid or a gas. *Hydrostatics* is the study of fluids at rest, such as the pressure of a fluid at a particular depth, or the *buoyant force* acting on an object in a fluid. *Pascal's principle* states that the pressure a fluid exerts on the walls of its container is constant at a given depth. *Archimedes' principle* states that the buoyant force acting on an object in a fluid is equal to the weight of the fluid displaced by the object.

Hydrodynamics is the study of fluids in motion. As a fluid flows through a pipe, the mass *flow rate* through the cross section is the same at any point in the pipe. *Bernoulli's equation* relates static pressure of a fluid to its dynamic (moving) pressure.

Key Terms

Archimedes' principle: the buoyant force acting on an object in a fluid is equal to the weight of the fluid displaced by the object

Bernoulli's principle: the sum of the pressures exerted by a fluid in a closed system is constant

density: the ratio of the mass to the volume of a substance

flow rate continuity: the volume or mass entering any point must also exit that point

fluid: any substance that flows, typically a liquid or a gas

hydrodynamics: the study of fluids in motion

hydrostatics: the study of fluids at rest

liquid: substance which has a fixed volume, but retains the shape of its container

KAPLAN 107

AP Physics
CHAPTER 6

pressure: force per unit area

Pascal: the SI unit for pressure equal to one newton of force per square meter of area

Pascal's principle: the pressure a fluid exerts on the walls of its container is constant at a given depth

Density and Pressure

The mass density ρ of a substance is the mass of the substance divided by the volume it occupies:

$$\rho = \frac{m}{V}$$

A *fluid* is any substance that flows and conforms to the boundaries of its container. A fluid could be a gas or a liquid. An *ideal fluid* is assumed to be incompressible (so that its density does not change), to flow at a steady rate, to be non-viscous (no friction between the fluid and the container through which it is flowing), and to flow irrotationally (no swirls or eddies).

Any fluid can exert a force perpendicular to its surface on the walls of its container. The force is described in terms of the pressure it exerts, or force per unit area:

$$p = \frac{F}{A}$$

The SI unit for pressure is the *Newton per meter squared*, or the *Pascal*. Sometimes pressure is measured in *atmospheres* (atm). One atmosphere is the average pressure exerted on us every day by the earth's atmosphere. The relationship between one atmosphere and Pascals is

$$1 \text{ atm} = 1.013 \times 10^5 \text{ Pa}$$

This is approximately equal to 15 lbs/in². In mechanics, it is often convenient to speak in terms of *mass* and *force*, whereas in fluids we often speak of *density* and *pressure*.

A static (non-moving) fluid produces a pressure within itself due to its own weight. This pressure increases with depth below the surface of the fluid. Consider a container of water with the surface exposed to the earth's atmosphere.

The pressure p_1 on the surface of the water is 1 atm, or 1.013×10^5 Pa. If we go down to a depth h from the surface, the pressure becomes greater by the product of the density of the water ρ, the acceleration due to gravity g, and the depth h. Thus the pressure p_2 at this depth is

$$p_2 = p_1 + \rho g h$$

Note that the pressure at any depth does not depend on the shape of the container, but rather only on the pressure at some reference level and the vertical distance below that level.

Pascal's Principle

Pascal's principle states that if we apply a change in pressure to a fluid which is completely enclosed in a container, the change in pressure is transmitted undiminished to every portion of the fluid and to the walls of the container. This is the principle behind many hydraulic devices, such as jacks or brakes. Consider a container filled with a fluid and containing a piston at either end. If we put pressure on one piston it will produce an equal pressure on the other piston, regardless of the relative sizes of the pistons.

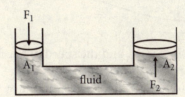

If we know the force applied to the piston on the left and the area of each piston, we can calculate the force on the piston on the right by setting the pressure on each piston equal to each other:

$$p_1 = p_2$$

$$\frac{F_1}{A_1} = \frac{F_2}{A_2}$$

$$F_2 = \frac{F_1 A_2}{A_1}$$

Archimedes' Principle

Archimedes' principle allows us to calculate the *buoyant force* acting on an object in a fluid. The buoyant force is the upward force exerted by the fluid on the object in the fluid, and is equal to the weight of the fluid which is displaced by the object. For example, if the floating object in the figure below displaces one liter of water, the buoyant force acting on the object is equal to the weight of one liter of water, which is about 10 N.

The buoyant force acting on an object in a fluid can be found by the equation

$$F_{buoyant} = \rho g V$$

where ρ is the density of the fluid, g is the acceleration due to gravity, and V is the volume of the displaced fluid. If the buoyant force is equal to the weight of the object in the fluid, the object will float.

Fluid Flow Continuity

Consider a fluid flowing through a tapered pipe:

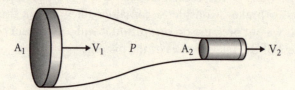

The area of the pipe on the left side is A_1, and the speed of the fluid passing through A_1 is v_1. As the pipe tapers to a smaller area A_2, the speed changes to v_2. Since mass must be conserved, the mass of the fluid passing through A_1 in a given amount of time must be the same as the mass of the fluid passing through A_2. If the density of the fluid is ρ_1, and the density of the fluid at A_2 is ρ_2, the *mass flow rate* through A_1 is $\rho_1 A_1 v_1$, and the mass flow rate through A_2 is $\rho_2 A_2 v_2$. Thus, by conservation of mass,

$$\rho_1 A_1 v_1 = \rho_2 A_2 v_2$$

This relationship is called the *equation of continuity*. If the density of the fluid is the same at all points in the pipe, the equation becomes

$$A_1 v_1 = A_2 v_2$$

The product of the area and the velocity of the fluid through the area is called the *volume flow rate*.

Bernoulli's Equation

Recall that in the absence of friction or other nonconservative forces, the total mechanical energy of a system remains constant, that is,

$$U_1 + K_1 = U_2 + K_2$$

$$mgy_1 + \frac{1}{2}mv_1^2 = mgy_2 + \frac{1}{2}mv_2^2$$

There is a similar law in the study of fluid flow, called *Bernoulli's principle*, which states that the total pressure of a fluid along any tube of flow remains constant. Consider a tube in which one end is at a height y_1 and the other end is at a height y_2:

FLUID MECHANICS AND THERMAL PHYSICS

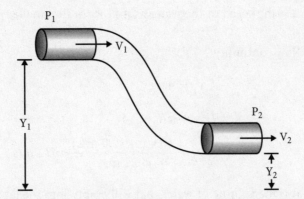

Let the pressure at y_1 be p_1 and speed of the fluid v_1. Similarly, let the pressure at y_2 be p_2 and the speed of the fluid v_2. If the density of the fluid ρ, Bernoulli's equation is:

$$p_1 + \frac{1}{2}\rho v_1^2 + \rho g y_1 = p_2 + \frac{1}{2}\rho v_2^2 + \rho g y_2$$

This equation states that the sum of the pressure at the surface of the tube, the dynamic pressure caused by the flow of the fluid, and the static pressure of the fluid due to its height above a reference level remains constant. Note that if we multiply Bernoulli's equation by volume, it becomes a statement of conservation of energy.

If a fluid moves through a horizontal pipe, the equation becomes:

$$p_1 + \frac{1}{2}\rho v_1^2 = p_2 + \frac{1}{2}\rho v_2^2$$

This equation implies that the higher the pressure at a point in a fluid, the slower the speed, and vice-versa.

Sample Problem:

A pipe carrying water at a speed of 2.0 m/s changes diameter from 0.80 meters to 0.20 meters as shown below. The smaller pipe then empties the water into the bottom of a freshwater lake that is 30 meters deep.

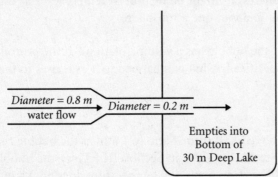

AP Physics
CHAPTER 6

(a) What is the speed of the water as it flows in the smaller pipe?

Answer:

Use the equation of flow continuity:

$$A_1 v_1 = A_2 v_2$$

$$\pi R_1^2 v_1 = \pi R_2^2 v_2$$

$$v_2 = \frac{(0.4 \text{ m})^2 (2.0 \text{ m/s})}{(0.1 \text{ m})^2} = 32 \text{ m/s}$$

(b) What is the volume of water that will empty into the lake each second?

Answer:

The water must leave the pipe at the rate: $\dfrac{\Delta Q}{\Delta t} = Av$.

In other words, the continuity equation above assures that the volume rate of flow is the same everywhere in the pipe and as it empties. Therefore, the volume (Q) of water that empties each second is Av for either section of the pipe.

$$Av = \pi R_1^2 v_1 = \pi R_2^2 v_2 = 1.00 \text{ m}^3/\text{sec}$$

(c) A bubble manages to get into the water as it flows through the pipe.
   i.   What is the pressure on the bubble when it exits the pipe into the bottom of the lake?

Answer:

The total pressure on the bubble is the pressure of the atmosphere plus the pressure of 30 m of water:

$$P_t = P_{atm} + \rho g h$$

$$= 1.013 \times 10^5 \text{ Pa} + (10^3 \text{ kg/m}^3)(9.8 \text{ m/s}^2)(30 \text{ m}) = 3.95 \times 10^5 \text{ Pa}$$

This is about 4 atmospheres. It might be helpful to remember that each 10 meter depth of fresh water is equivalent to about one atmosphere.

   ii. If the bubble has a volume of 10 ml at the bottom of the lake, specifically what will happen to it as it rises to the top of the lake?

Answer:

Since the pressure will be reduced from 4 atm to 1 atm as the bubble rises, its volume will be increased by 4 times. (See $P_1 V_1 = P_2 V_2$ in Section III.) Thus, the bubble will have a volume of 40 ml as it reaches the surface, assuming temperature does not change appreciably.

112   **KAPLAN**

Physics Review

FLUID MECHANICS AND THERMAL PHYSICS

II. TEMPERATURE AND HEAT

The total internal energy of the molecules of a substance is called *thermal energy*. The *temperature* of a substance is a measure of the average kinetic energy of the molecules in the substance, and gives an indication of how hot or cold the substance is relative to some standard. The energy transferred between two substances because of a temperature difference is called *heat*. Heat can be transferred by *conduction, convection*, or *radiation*. The amount of heat required to raise the temperature of a certain amount of substance is called *specific heat*. Many substances expand when heated. When a substance gains or loses a certain amount of heat without changing temperature, at its freezing point or boiling point, it undergoes a *phase change*, (for example, changing from a solid to liquid or a liquid to a gas.)

Key Terms

absolute zero: the lowest possible temperature, at which all molecular motion would cease and a gas would have no volume.

calorie: the amount of heat required to raise the temperature of one gram of water by one Celsius degree

calorimeter: device which isolates objects to measure temperature changes due to heat flow

Celsius (C): temperature scale in which the freezing point of water is 0 and the boiling point of water is 100

convection: heat transfer by the movement of a heated substance, due to differences in density

conduction: heat transfer from molecule to molecule in substances due to difference in temperature

conductor: a material through which heat or electric current can easily flow

heat: the energy which is transferred from one body to another due to a temperature difference

heat of fusion: energy needed to change a unit mass of a substance from a solid to a liquid state at the melting point

heat of vaporization: energy needed to change a unit mass of a substance from a liquid to a gaseous state at the boiling point

joule heating: the increase in temperature in an electrical conductor due to the conversion of electrical energy into heat energy

Joule's law of heating: the heating power of an electric current through a resistance is equal to the product of the current and the voltage across the resistor

KAPLAN 113

Kelvin (absolute) temperature scale: scale in which zero Kelvins is defined as absolute zero, the temperature at which all molecular motion ceases

melting point: the temperature at which a substance changes from a solid to a liquid state

phase change: the process of a substance changing from one state, such as a solid, liquid, or gas, to another state

radiation: the transmission of energy by electromagnetic waves

specific heat capacity: the quantity of heat required to raise the temperature of one mass unit of a substance by one degree

temperature: the property of a body which indicates how hot or cold a substance is with respect to a standard

thermal energy: the sum of the internal potential and kinetic energy of the random motion of the molecules making up an object

thermal equilibrium: state between two or more objects in which temperature doesn't change

thermal expansion: increase in length or volume of a material due to an increase in temperature

Heat and Temperature

There are three states of matter we will be discussing: solid, liquid, and gas. The molecules of a solid are fixed in a rigid structure. The molecules of a liquid are loosely bound and may mix with one another freely. Also, while the liquid has a definite volume, it still takes the shape of its container. The molecules of a gas interact with each other only slightly, and generally move at high speeds compared with the molecules of a liquid or a solid. But in the case of all three states of matter, the molecules are moving and therefore have energy. They have potential energy because of the bonds between them and kinetic energy because the molecules have mass and speed. Relatively speaking, the potential energy between gas molecules can be ignored, and later we will focus only on their kinetic energy. The sum of the potential and kinetic energies of the molecules in a substance is called the *internal energy* of the substance. When a warmer substance comes in contact with a cooler substance, some of the kinetic energy of the molecules in the warmer substance is transferred to the cooler substance. The energy representing the kinetic energy of molecules that is transferred spontaneously from a warmer substance to a cooler substance is called *heat energy*.

Temperature is the measure of how hot or cold a substance is relative to some standard. It is the measure of the *average kinetic energy* of the molecules in a substance. The two temperature scales used most widely in scientific applications are the Celsius scale and the Kelvin scale. The only difference between them is where each starts. On the Celsius scale, the freezing point of water is 0° C, and the boiling point of water (at standard pressure) is 100° C.

The Kelvin scale has temperature units which are equal in size to the Celsius degrees, but the temperature of 0 Kelvin is *absolute zero*, defined as the temperature at which all molecular motion in a substance ceases. Zero Kelvin is equal to $-273.15°$ C, so we can convert between the Kelvin scale and the Celsius scale by the equation

$$K = °C + 273$$

Note that we have rounded 273.15 to 273. The boiling point of water in Kelvins would be $K = 100°$ C + 273 = 373 K.

Mechanical Equivalent of Heat

The unit most often used for mechanical energy is the *joule*. Historically, the unit for heat has been the *calorie*. One calorie is defined as the heat needed to raise the temperature of one gram of water by one degree Celsius. In the mid-19th century, the British engineer James Joule showed that there is a relationship between energy in the form of work in joules and energy in the form of heat in calories. Joule performed experiments which revealed that doing mechanical work on a substance can make its temperature rise. For example, rubbing your hands together causes them to heat up, or stirring a drink adds heat to it. The conversion between joules and calories is

$$1 \text{ calorie} = 4.186 \text{ joules}$$

The numbers here are not really important for the AP Physics B exam, but you should understand the concept that mechanical work and heat are both forms of energy and can be converted into one another.

Thermal Expansion of a Solid

When a solid is heated, it typically expands. Different substances expand at different rates, which is why it makes sense to heat the lid of a glass jar when it is too tight. When it is heated, the metal lid will expand more than the glass jar, making it easier to loosen. Solids undergo two types of expansion when heated: *linear thermal expansion*, which is the increase in any one dimension of the solid, and *volume thermal expansion*, which results in an increase in the volume of the solid. In the case of linear expansion, the change in length Δl is proportional to the original length l_0 and the change in temperature ΔT of the solid:

$$\Delta l = \alpha l_0 \Delta T$$

where α is the coefficient of linear expansion.

The same is true for volume expansion. The change in volume ΔV is proportional to the original volume of the solid and its change in temperature ΔT:

$$\Delta V = \beta V_0 \Delta T$$

where β is the coefficient of volume expansion.

AP Physics

CHAPTER 6

Heat Transfer

There are three was of transferring heat from one place to another:

Conduction. Conduction is the transfer of heat directly through a material, or by actual contact between two materials. Metals are typically good heat *conductors*. In fact, materials which are good electrical conductors are usually good heat conductors as well. A material which is not a good heat conductor, like wood or air, is called an *insulator*. If you place an iron skillet on a fire, heat is transferred by conduction to the handle of the skillet. If you grasp the iron handle with your bare hand, you will feel it transfer heat to your hand by conduction.

Convection. Convection is the transfer of heat by the bulk movement of a fluid. If the air near the floor of a cool room is heated, it expands and becomes less dense than the air above it, causing it to rise. As it rises, it cools, becomes more dense again, and falls toward the floor. If the air near the floor is continually heated, the cycle will repeat itself. Water heated in a pan is an example of heat transfer by convection, since water near the bottom of the pan near the fire is heated, rises, cools, then falls again. If the temperature gets high enough, the water begins to boil as it cools itself by transferring heat to the air by convection.

Radiation. Radiation is the process by which heat is transferred by electromagnetic waves. We receive heat from the sun by radiation, principally in the form of light, infrared, and ultraviolet waves. Microwave ovens use microwaves to transfer heat to food. If you stand near a roaring campfire, you will feel the heat radiating from the fire in the form of light and infrared rays.

Specific Heat and Heat Capacity

Heat is supplied to (or absorbed by) a system to raise its temperature; conversely, heat is released if it cools. The heat absorbed or released by an object as a result of a change in temperature is calculated from the equation:

$$Q = mc\Delta T$$

where Q is the symbol for heat, m is the mass of the object, ΔT is the change in temperature and is equal to the final temperature minus the initial temperature, and c is a quantity known as the *specific heat* of the substance.

The more massive a substance is, the more heat is required to bring about a particular change in temperature. Recall that the temperature of a substance is a measure of the average kinetic energy of the particles in the substance. This applies to all states of matter. In solids, for example, the atoms vibrate about their equilibrium positions; the stronger these vibrations, the higher the temperature of the solid. A certain amount of heat supplied to a large number of particles would not increase their average energy by much; however, if there were only a small number of particles in the system, that same amount of heat is now not spread as thin, and thus each particle would gain a larger amount of energy, bringing up the temperature more. The more of a substance there is, the more heat is required to bring about a change in its temperature.

116 **KAPLAN**

Physics Review

FLUID MECHANICS AND THERMAL PHYSICS

Yet, not every substance is responsive to heat to the same degree. Even though we expect that to raise 2 kg of a substance by 1 C° requires more heat than raising 1 kg of the same substance by 1 C° (in fact, it requires twice the amount of heat), we would not expect that the same amount of heat is required to raise the temperature of 1 kg of steel versus 1 kg of plastic. The *specific heat c* is a proportionality constant that gives an indication of the ease with which one can raise the temperature of something; the larger it is, the larger the amount of heat required to raise its temperature a certain number of degrees, and also the more heat is released if it cools by a certain number of degrees. Its value is a property of the nature of the substance and does not change based on the amount of stuff we have, since that has already been taken into account by the mass. The specific heat is often more formally defined as the heat necessary to raise the temperature of 1 kg or 1 g of a material by 1 C° or 1 K. Iron, for example, has a specific heat of about 0.1 kilocalorie/kg C°, while water has a specific heat of 1.0 kcal/kg C°. It is therefore much easier to raise the temperature of 1 kg of iron by 10 C° than it is to do the same to 1 kg of water. In fact, you should be able to see that ten times the heat is needed.

For example, which would take more heat to raise its temperature by 1 C°: 1 kg of aluminum or 1 kg of wood? The 1 kg piece of wood would take more heat, since its specific heat is higher. It doesn't take much heat to make aluminum feel hot, but it takes more heat to make a piece of wood feel hot.

Mass and specific heat are sometimes lumped together to give a quantity known as the heat capacity. This quantity describes the heat needed to raise the temperature of the object as a whole by 1 C° or 1 K.

Phase Changes

While heat is associated with the change of thermal energy, a system does not necessarily increase in temperature. Heat can also increase the potential (rather than kinetic) energy of the particles in a system; this occurs during a *phase change*. Heat is required to melt something (change its phase from a solid to a liquid), or to vaporize something (change its phase from a liquid to a gas), or to sublimate something (change its phase directly from a solid to a gas). In all these cases, the molecules are overcoming the attractive forces that hold them together. This is where the energy supplied by heating is being put to use. Conversely, heat is released as a substance freezes or condenses. During such phase changes the temperature remains constant, and the heat involved in these processes can be expressed as:

$$Q = mL$$

where m is the mass of the substance undergoing the phase change and L is the heat of transformation, the value of which depends on both the substance and the particular process we are talking about: vaporization, sublimation, or fusion (melting).

For example, the heat of fusion L_f of ice is 3.3×10^5 joules/kg and the specific heat of water is 4.2×10^3 joules/kg C°. How much heat is required to completely melt 10 kg of ice and then raise the temperature of the water from 0° C to 30° C?

KAPLAN 117

The heat required to melt the ice and then raise the temperature of the water is

$$Q = mL_f + mc\Delta T$$

$$Q = (10 \text{ kg})(3.3 \times 10^5 \text{ joules/kg}) + (10 \text{ kg})(4.2 \times 10^3 \text{ joules/kg C°})(30 - 0)$$

$$Q = 4.56 \times 10^6 \text{ joules.}$$

This is approximately one million calories.

If we were to continue to heat the water until it boiled and turned to steam at 100° C, and then continued to add heat to the steam to raise its temperature beyond 100° C, the temperature would not change during the phase changes from ice to water and water to steam. The heat added would go into changing the phase of the water, not its temperature. A graph of temperature vs. heat added for this process is shown below:

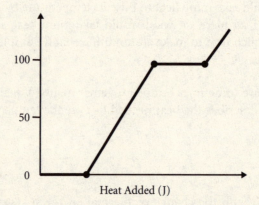

The flat portions imply that the temperature is not changing, and correspond to the processes of melting and boiling.

III. KINETIC THEORY AND THERMODYNAMICS

Kinetic molecular theory involves the study of matter, particularly gases, as very small particles in constant motion. We study gases by relating their *pressure, volume*, and *temperature* in the *ideal gas law*. The laws that govern the flow of heat in or out of a system are called the *first and second laws of thermodynamics*. *Thermodynamics* is the study of heat transfer. Often we analyze the energy transfer of a system using a *pressure-volume (PV) diagram*.

Key Terms

adiabatic: the expansion or compression of a gas without a gain or loss of heat

Carnot efficiency: the ideal efficiency of a heat engine or refrigerator working between two constant temperatures

Physics Review

FLUID MECHANICS AND THERMAL PHYSICS

entropy: the measure of the amount of disorder in a system

first law of thermodynamics: the heat lost or gained by a system is equal to change in internal energy of the system minus work done on the system; conservation of energy

heat engine: device which changes internal energy into mechanical work

ideal gas law: the law which relates the pressure, volume, number of moles, and temperature of an ideal gas

isobaric: any process in which the pressure of a gas remains constant

isochoric (or *isovolumetric*): any process in which the volume of a gas remains constant

isothermal: any process in which the temperature of a gas remains constant

kinetic molecular theory: the description of matter as being made up of extremely small particles which are in constant motion

mole: one mole of a substance contains Avogadro's number (6.02×10^{23}) of molecules or atoms

pressure-volume (PV) diagram: a graph of pressure vs. volume which gives an indication of the work done by or on a system, and the energy transferred during a process

second law of thermodynamics: heat flows naturally from a region of higher temperature to a lower temperature; all natural systems tend toward a state of higher disorder

thermodynamics: the study of heat transfer

Thermodynamics

Thermodynamics is the study of heat transfer. There are actually four laws of thermodynamics, but we will look at two of them which are required for the AP Physics B exam. First we should define some terms.

1. *Isolated.* A system is said to be isolated when it cannot exchange energy or matter with its surroundings, like a well-insulated thermos flask.
2. *Closed.* A system is said to be closed when it can exchange energy but not matter with the surroundings, like a test tube with a stopper in it.
3. *Open.* A system is said to be open when it can exchange both matter and energy with the surroundings, as with a pot of boiling water allowing water vapor to escape into the air.

The First Law of Thermodynamics

As we've discussed in previous chapters, energy can be transformed in many forms, but is conserved; that is, the total amount of energy must remain constant. This is true of a system

KAPLAN 119

only if it is isolated. Since energy can neither go in nor go out, it has to be conserved. If the system is closed or open, the amount of energy in the system can certainly change. A system can exchange energy with its surroundings in two general ways: as heat or as work. The first law of thermodynamics states that *the change in the internal energy ΔU of a system is equal to the heat Q lost or gained by the system plus the work W done on the system*:

$$\Delta U = Q + W$$

On the AP Physics B exam, if work is done ON a system, W is *positive*. Note that in some textbooks W is defined as the work done BY, rather than ON the system, in which case the equation is written as $\Delta U = Q - W$, and the work done BY the system is considered positive. Regardless of which convention is used, if work is done on a system, its energy would increase. If work is done by the system, its energy would decrease. Work is generally associated with movement against some force. For ideal gas systems, for example, *expansion* against some external pressure means that work is done by the system, while *compression* implies work being done on the system.

For example, if a system has 50 J of heat added to it, resulting in 20 J of work being done by the system, the change in internal energy of the system is $\Delta U = Q + W = 50 \text{ J} + (-20 \text{ J}) = 30 \text{ J}$.

If heat is added to a system and no work is done, then the heat lost by one element in the system is equal to the heat gained by another element. For example, if a sample of metal of mass m_m and specific heat c_m is heated and then dropped into a beaker containing a mass m_w of water, then the final temperature of the water can be found by:

$$Q \text{ lost by the metal} + Q \text{ gained by the water} = 0$$
$$m_m c_m (T_f - T_i)_m + m_w c_w (T_f - T_i)_w = 0$$

Since the final equilibrium temperature is the same for the metal and the water, we can solve for the final temperature T_f. This demonstrates conservation of energy, since there was no heat exchange with the environment.

Heat Engine

A heat engine is any device that uses heat to perform work. There are three essential features of a heat engine:

1. Heat is supplied to the engine at a high temperature from a hot reservoir.

2. Part of the input heat is used to perform work.

3. The remainder of the input heat which did not do work is exhausted into a cold reservoir, which is at a lower temperature than the hot reservoir.

FLUID MECHANICS AND THERMAL PHYSICS

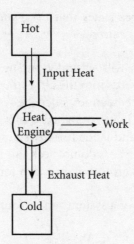

The *percent efficiency* %e of the heat engine is equal to the ratio of the work done to the amount of input heat:

$$\%e = \frac{Work}{Q_{Hot}} \times 100$$

For example, a heat engine extracts 100 J of energy from a hot reservoir, does work, then exhausts 40 J of energy into a cold reservoir. The work done is equal to the difference between the input heat and the output heat:

$$W = 100 \text{ J} - 40 \text{ J} = 60 \text{ J}$$

The percent efficiency is $\frac{60 \text{ J}}{100 \text{ J}} = 60\%$

The French engineer Sadi Carnot suggested that a heat engine has maximum efficiency when the processes within the engine are reversible, that is, both the system and its environment can be returned to exactly the states they were in before the process occurred. In other words, there can be no dissipative forces, like friction, involved in the *Carnot cycle* of an engine for it to operate at maximum efficiency. All spontaneous processes, such as heat flowing from a hot reservoir to a cold reservoir, are not reversible, since work would have to be done to force the heat back to the hot reservoir from the cold reservoir (a refrigerator), thus changing the environment by using some of its energy to do work. A reversible engine is called a *Carnot engine*.

The Second Law of Thermodynamics

Entropy S is a measure of the disorder, or randomness, of a system. The greater the disorder of a system, the greater the entropy. If a system is highly ordered, like the particles in a solid, we say that the entropy is low. At any given temperature, a solid will have a lower entropy than a gas, because individual molecules in the gaseous state are moving randomly, while individual molecules in a solid are constrained in place. Entropy is important because it determines whether a process will occur spontaneously.

CHAPTER 6

The second law of thermodynamics states that *all spontaneous processes proceeding in an isolated system lead to an increase in entropy.*

In other words, an isolated system will naturally pursue a state of higher disorder. If you watch a magician throw a deck of cards into the air, you would expect the cards to fall to the floor around him in a very disorderly manner, since the system of cards would naturally tend toward a state of higher disorder. If you watched a film of a magician, and his randomly placed cards jumped off the floor and landed neatly stacked in his hand, you would believe the film is running backward, since cards do not seek this state of order by themselves. Thus, the second law of thermodynamics gives us a direction for the passage of time.

An increase or decrease in entropy of a system can be found by:

$$\Delta S = \frac{Q}{T}$$

where Q is heat flow into or out of a system and T is the average Kelvin temperature over which the system changes. [Note: Using average temperature is an approximation for an actual integration using the method of calculus, but the result is close to the actual.]

Ideal Gases

Among the different phases of matter, the gaseous phase is the simplest to understand and to model, since all gases display similar behavior and follow similar laws regardless of their identity. The atoms or molecules in a gaseous sample move rapidly and are far apart. In addition, intermolecular forces between gas particles tend to be weak; this results in certain characteristic physical properties, such as the ability to expand to fill any volume and to take on the shape of a container. Furthermore, gases are easily, though not infinitely, compressible.

The state of a gaseous sample is generally defined by four variables: pressure (p), volume (V), temperature (T), and number of moles (n), though as we shall see, these are not all independent. A *mole* of a substance is Avogadro's number (6.02×10^{23}) of molecules or atoms of that substance. The *pressure* of a gas is the force per unit area that the atoms or molecules exert on the walls of the container through collisions. The SI unit for pressure is the *pascal* (Pa), which is equal to one newton per meter squared. Sometimes gas pressures are expressed in atmospheres (atm). One atmosphere is equal to about 10^5 Pa, and is equal to the average pressure the earth's atmosphere exerts on us each day. Volume can be expressed in liters (L) or cubic meters (m^3), and temperature is measured in Kelvins (K) for the purpose of the gas laws. Recall that we can find the temperature in K by adding 273 to the temperature in Celsius. Gases are often discussed in terms of standard temperature and pressure (STP), which refers to the conditions of a temperature of 273 K (0°C) and a pressure of 1 atm.

When examining the behavior of gases under varying conditions of temperature and pressure, it is most convenient to treat them as *ideal gases*. An ideal gas represents a hypothetical gas whose molecules have no intermolecular forces; that is, they do not interact with each other, and they occupy no volume. Although gases in reality deviate from this idealized behavior, at relatively low pressures and high temperatures many gases behave in nearly ideal fashion. Therefore, the assumptions used for ideal gases can be applied to real gases with reasonable accuracy.

The Ideal Gas Law

The ideal gas law gives the relationship between the pressure, volume, temperature, and number of moles of a gas before and after some process:

$$pV = nRT$$

where R is the universal gas constant, equal to 8.31 J/(mol K).

If the number of moles of a gas remains constant during a process, the product of pressure and volume divided by temperature remains constant:

$$\frac{P_1 V_1}{T_1} = \frac{P_2 V_2}{T_2}$$

If temperature remains constant during a process (isothermic), the equation becomes

$$P_1 V_1 = P_2 V_2 \text{ (Boyle's Law)}$$

If the volume remains constant during a process (isochoric), the equation becomes

$$\frac{P_1}{T_1} = \frac{P_2}{T_2}$$

If the pressure remains constant during a process (isobaric), the equation becomes

$$\frac{V_1}{T_1} = \frac{V_2}{T_2} \text{ (Charles law)}$$

Pressure—Volume (pV) Diagrams

We can study the changes in pressure, volume, and temperature of a gas by plotting a graph of pressure vs. volume for a particular process. We call this graph a pV diagram. For example, let's say that a gas starts out at a pressure of 4 atm and a volume of 2 liters, as shown by the point A in the pV diagram below:

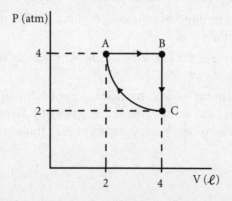

AP Physics

CHAPTER 6

If the pressure of the gas remains constant but the volume changes to 4 liters, then we trace a line from point *A* to point *B*. Since the pressure remains constant from *A* to *B*, we say that the process is *isobaric*. If we decrease the pressure to 2 atm but keep the volume constant, we trace a line from *B* to *C*. This constant-volume process is *isochoric*, or *isovolumetric*. If we want the gas to return to its original state without changing temperature, we must trace a curve from point *C* to *A* along an *isotherm*. Note that an isotherm on a *pV* diagram is not a straight line. The work done during the process *ABCA* is the area enclosed by the graph, since $W = -p\Delta V$. In this case the work done **on** the system is positive.

Any process which is done without the transfer of heat is called an *adiabatic* process. Since there is no heat lost or gained in an adiabatic process, then the first law of thermodynamics states that the change in internal energy of a system is simply equal to the work done on or by the system, that is, $\Delta U = W$.

The change in internal energy ΔU of a gas during an adiabatic process is given by the equation:

$$\Delta U = nc_v\Delta T$$

where *n* is the number of moles of the gas and c_v is the molar specific heat of the gas at constant volume. At constant pressure, we replace the molar specific heat at constant volume with the molar specific heat at constant pressure, c_p.

The Kinetic Theory of Gases

As indicated by the gas laws, all gases show similar physical characteristics and behavior. A theoretical model to explain why gases behave the way they do was developed during the second half of the 19th century. The combined efforts of Boltzmann, Maxwell, and others led to the kinetic theory of gases, which gives us an understanding of gaseous behavior on a microscopic, molecular level. Like the gas laws, this theory was developed in reference to ideal gases, although it can be applied with reasonable accuracy to real gases as well.

The assumptions of the kinetic theory of gases are as follows:

1. Gases are made up of particles whose volumes are negligible compared to the container volume.

2. Gas atoms or molecules exhibit no intermolecular attractions or repulsions.

3. Gas particles are in continuous, random motion, undergoing collisions with other particles and the container walls.

4. Collisions between any two gas particles are elastic, meaning that no energy is dissipated and kinetic energy is conserved.

5. The average kinetic energy of gas particles is proportional to the absolute (Kelvin) temperature of the gas, and is the same for all gases at a given temperature. As listed in the list of equations, the average kinetic energy of each molecule is related to temperature T by the equation:

124 **KAPLAN**

FLUID MECHANICS AND THERMAL PHYSICS

$K_{avg} = \dfrac{3}{2}k_B T$, where k_B is the Boltzmann constant, 1.38×10^{-23} J/K. The root-mean-square speed of each molecule can be found by $v_{rms} = \sqrt{\dfrac{3k_B T}{\mu}}$, where μ is the mass of each molecule.

Sample Problem:

Consider a part of the Universe in which two very large systems at temperatures of 200 and 300 Kelvins come into contact with each other. A small amount of heat, 40 Joules, flows from System 1 to System 2.

(a) Calculate the change in entropy of System 1.

Answer:

System 1 (originally at lower temperature):

$$\Delta S = \frac{Q}{T_{K(ave)}} = \frac{(40\ J)}{(200\ K)} = 0.20\ J/K$$

Though it's an oversimplification, entropy change can be calculated by using the heat flow into or out of the system divided by the *average Kelvin* temperature over which the entropy change occurs. Since these two systems are quite large, we can assume that the temperatures do not change by much, so we use the temperatures given. We can also assume that heat will naturally flow from the higher to lower temperature, so Q is positive in the above case.

(b) Calculate the change in entropy of System 2.

Answer:

System 2 (originally at higher temperature):

$$\Delta S = \frac{(-40\ J)}{(300\ K)} = -0.13\ J/K$$

(c) Calculate the net change in entropy of the Universe due to the heat flow above.

Answer:

$$\Delta S = S_1 + S_2 = 0.20\ J/K - 0.13\ J/K = +0.07\ J/K$$

(d) Briefly discuss conservation of energy and conservation of entropy in this process.

Answer:

Even though the same amount of heat flowed from one system into the other, when we add the two to get the net change in entropy, the net change is +0.07 J/K, a net positive entropy change in the Universe as a result of this heat flow. Energy is conserved, but entropy increases.

KAPLAN 125

AP Physics
CHAPTER 6

Important Equations

Conversion from Celsius to Fahrenheit:	$T_F = (9/5)T_C + 32$
Conversion from Fahrenheit to Celsius:	$T_C = (5/9)(T_F - 32)$
Two forms of the ideal gas law (molecular/molar):	$pV = NkT = nRT$
Conversion from Celsius to Kelvin:	$T_K = T_C + 273.15$
Root mean square velocity of molecules:	$v_{rms} = (3RT/M)^{1/2} = (3k_B T/\mu)^{1/2}$
Statement of equivalence of internal energy of a gas to the average kinetic energies of its molecules:	$K_{ave} = (3/2)kT = (1/2)m<v>^2$
Heat required to change temperature:	$Q = mc\Delta T$
Heat of combustion:	$Q = mH$
Latent heat of fusion/vaporization:	$Q = mL$
First Law of Thermodynamics:	$\Delta U = Q + W$
Work in an isobaric process:	$W = -P\Delta V$
Change in internal energy for adiabatic process:	$\Delta U = nC_v\Delta T$
Efficiency of an engine:	$\varepsilon = \dfrac{|W|}{|Q_H|}$
Efficiency of a Carnot engine:	$\varepsilon_{Carnot} = \dfrac{(T_H - T_L)}{T_H}$
Entropy:	$\Delta S = \dfrac{\Delta Q}{T}$
Total pressure in a fluid:	$p = p_o + \rho g h$
Buoyant force in a fluid:	$F_B = \rho V g$
Fluid equation of continuity:	$A_1 v_1 = A_2 v_2$
Bernoulli's equation:	$p + \rho g y + \dfrac{1}{2}\rho v^2 = constant$
Pressure:	$P = F/A$
Change in length due to temperature change:	$\Delta L = \alpha L_o \Delta T$

126 KAPLAN

PRACTICE PROBLEMS

1. A 150 g bullet made of ice at 0°C is shot from a gun and travels at 500 m/s before imbedding itself in a block of wood.
   (a) If the bullet's kinetic energy is totally converted to heat that it also absorbs, how much ice melts?
   (b) Take that same energy and assume that five moles of an ideal gas at room temperature (20°C) do that much work while expanding at constant pressure.
      i. What is the change in volume of the gas?
      ii. What is the final temperature of the gas?
   (c) What is the Carnot efficiency of an engine operating between these two temperatures?

2. An ideal gas is taken through the process *A-B-C-D-A* as shown below.

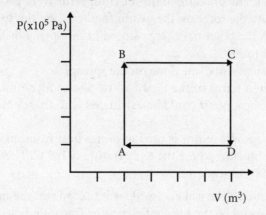

   (a) Which process (if any) is isobaric? Justify your answer.
   (b) Which process (if any) is isothermal? Justify your answer.
   (c) Which process (if any) is isochoric? Justify your answer.
   (d) Which process (if any) is adiabatic? Justify your answer.
   (e) Find the net work done by the gas or on the gas (specify which) in the complete cycle.
   (f) What is the value of Q for the entire problem?

AP Physics
CHAPTER 6

3. A piston of air is used to compress a spring, as shown below. In piston **A**, the air in the upper chamber has a volume of 0.002 m³ at standard temperature and atmospheric pressure. In piston **B**, the air in the upper chamber has been heated by combustion to 177° C, increasing the volume to 0.003 m³ and compressing the spring. The cross-sectional area of each piston is 0.015 m². The spring constant of the spring is 2×10^4 N/m.

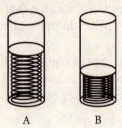

A   B

(a) Calculate the pressure in the air compartment in piston *B*.
(b) Calculate the force on the piston (and thus on the spring) in *B*.
(c) What is the potential energy stored in the spring by the expansion from *A* to *B*?
(d) How much work was done on the spring?
(e) Explain in terms of the First Law of Thermodynamics how the specific heat of air could be calculated with this demonstration.

4. A 0.250 horsepower motor is used to pump heat from inside a room at 30° C to the outside, where the temperature is 10° C. The room has a volume of 600 cubic meters.

(a) How much electrical energy does it take to run the motor each hour?
(b) What would be the ideal efficiency, or Carnot efficiency, of a heat engine that operates between these temperatures?
(c) If the actual efficiency of the motor is 75%,
  i. how much useful work does the motor do each hour?
  ii. how much can the heat released by the motor actually raise the temperature of the air in the room in which it is operating each hour? (Use 0.733 J/g for specific heat of air and 1.29 kg/m³ for density of air.)
(d) If the motor is disconnected and taken outside, by how much will an iron rod in the motor that is 4.5 cm long change in length? (The coefficient of linear expansion for iron is $12 \times 10^{-6}/C°$.)

FLUID MECHANICS AND THERMAL PHYSICS

5. Let's suppose a relaxed swimmer with mass 55 kg floats with 90% of his body under water.

   (a) What is the volume of the person?
   (b) What is the specific gravity of the person?
   (c) What is the buoyant force of the person when he is floating?

   Now the swimmer decides to swim to the bottom of a deep lake, to a depth of 40 meters.

   (d) What is the total pressure on the swimmer at this depth?
   (e) Assuming the human body to be somewhat compressible, what is the effect of this change in pressure on:
       i. the density of the body?
       ii. the buoyant force on the body?

6. Water flows in through a 3-cm diameter pipe in the basement level of a house at a rate of 12.5 m/s to an open second floor faucet. The second floor faucet is located 7.2 meters above the basement inlet pipe. Assume all the pipes in the house have the same diameter.

   (a) What will be the difference in water pressure between the basement inlet pipe and the pipes on the second floor when the water is not running?
   (b) What is the volume rate of flow of water into the house?
   (c) The faucet valve on the second floor has 1/2 the diameter of the water pipes. What is the volume rate of flow of water out of the faucet?
   (d) What will be the rate of flow (in m/s) out of the faucet valve?

AP Physics

CHAPTER 6

Solutions to Practice Problems

1. Answers:

(a) The total energy of the bullet is: $KE = \frac{1}{2}mv^2 = \frac{1}{2}(0.15)(500)^2 = 18{,}750$ J

A mass of 0.150 kg of ice requires $mL_f = (0.15)(335{,}000) = 50{,}250$ J to melt it all, so we know that there will be some ice remaining. How much?

18,750 J = m(335,000 J/kg). With $m = 0.0559$ kg, we see that 55.9 g of the ice melts.

(b) i. In an isobaric process, $W = P\Delta V$. The work = 18,750 J and atmospheric pressure = 1.013×10^5 Pa.

Thus, $\Delta V = \dfrac{18{,}750 \text{ J}}{1.013 \times 10^5 \text{ Pa}} = 0.185$ m^3

(b) ii.

From the ideal gas law, $P\Delta V = nR\Delta T$

$(1.013 \times 10^5)(0.185) = (5)(8.31)(\Delta T)$

and $\Delta T = 451$ C° = 451 K

Since the initial temperature was 20°C, the final temperature is 471°C = 744 K.

(c) Carnot efficiency depends on the Kelvin temperature of the hot and cold reservoirs.

$$\varepsilon_{\text{Carnot}} = \frac{(T_H - T_L)}{T_H} = \frac{(744 - 293)}{744} = 0.606 = 60.6\%$$

2. Answers:

(a) An isobaric process is one undergone at constant pressure, which would be both steps BC and DA.

(b) An isothermal process is one undergone at constant temperature. None of the steps are isothermal, since either pressure or volume changes in each step while the other does not change. By $PV = nRT$, a change in either P or V alone cannot happen without a joint change in T. However, the entire cycle is isothermal, since it returns to original conditions.

(c) An isochoric process is one at constant volume, also called isovolumetric. Steps AB and CD occur at constant volume.

(d) An adiabatic process is one that occurs without any heat flow in or out of the system. None of the steps are adiabatic.

(e) As shown in the diagram, the work done by the gas during the cycle is the area inside the cycle, shown as the shaded area in the diagram below. The numerical value for work is calculated as the area of the rectangle enclosed here, which is:

130   **KAPLAN**

$W = \Delta P \Delta V = (4 \times 10^5 \text{ Pa} - 0.5 \times 10^5 \text{ Pa})(6 \text{ m}^3 - 2 \text{ m}^3) = 1.4 \times 10^6 \text{ J}$

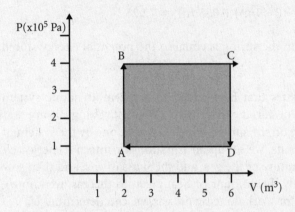

(f) We use $Q = U - W$ (wherein positive work is defined as work done *on* the system). For the complete cycle, there is no net change in temperature, since it returns to original conditions at A. No change in temperature indicates no net change in internal energy, U. Therefore, $Q = -W$. We have already calculated the net work done by the system in part (e). Since Q equals $-W$, the heat exchange is 1.4×10^6 Joules. The net work is done by the molecules, or more work is done in expansion than in compression. The net work is negative by our definition of work, so the Q is positive, meaning heat is absorbed. The heat put into the system is used by the molecules to do the work of expansion.

3. Answers:

(a) Use the combined gas law: $\dfrac{P_1 V_1}{T_1} = \dfrac{P_2 V_2}{T_2}$

Remember that temperatures must be converted to Kelvins.

$$\frac{(1 \text{ atm})(0.002 \text{ m}^3)}{273 \text{ K}} = \frac{(P_2)(0.003 \text{ m}^3)}{450 \text{ K}}$$

$P_2 = 1.1$ atm or 1.114×10^5 Pa

Note the conversion, using 1 atm = 1.013×10^5 Pa

(b) Use the equation for pressure to calculate force: $P = \dfrac{F}{A}$

$$F = PA = (1.114 \times 10^5 \text{ Pa})(0.015 \text{ m}^2) = 1{,}700 \text{ N}$$

(c) First, using Hooke's Law to determine how much the spring was compressed:

$F = kx$

$x = \dfrac{F}{k} = \dfrac{(1{,}700 \text{ N})}{(2 \times 10^4 \text{ N/m})} = 0.085 \text{ m}$

AP Physics
CHAPTER 6

Now, use the equation for spring potential energy:

$$U_s = \frac{1}{2}kx^2 = \frac{1}{2}(2 \times 10^4 \text{ N/m})(0.085 \text{ m})^2 = 72.25 \text{ J}$$

(d) The work done on the spring is equal to the potential energy stored in the spring, 72.25 Joules.

(e) The First Law states that heat added to a thermodynamic system is used to increase internal energy and/or for the system to do work. Using an apparatus like this, we can determine work done on the spring, which is work done by the system of gas molecules. With the proper gas constants, we set up an equation for internal energy, U, of the gas, knowing the change in temperature of the gas, and the mass of gas, and then solve for specific heat. If we calibrate separately the amount of heat given to the gas, we know Q. Since $Q = U - W$, where W is negative for work done by the gas, we can determine U.

4. Answers:

(a) An engine requiring 0.250 hp to run consumes

$(0.25 \ hp)(746 \text{ W/1 hp}) = 186.5 \text{ W} = 186.5 \text{ J/s of energy.}$

In one hour, that makes 671,400 J of energy input to the engine.

(b) Remember that Kelvin temperatures must be used in this equation.

$$\text{Carnot efficiency} = \frac{1 - T_c}{T_H} = 1 - \frac{283 \text{ K}}{303 \text{ K}} = 0.066 \text{ or } 6.6\%$$

(c) i. An efficiency of 75% means that 25% of that input energy will be produced as heat (instead of useful work) by the motor. So,

$(0.25)(671,400 \text{ J}) = 167,850 \text{ J}$ of heat produced per hour.

(c) ii. Use the formula $Q = mc\Delta T$, with $Q = 167,850$ J, $c = 0.733$ J/g, and $\rho = 1.29$ kg/m^3

Calculate the mass of air in the room using the density formula:

$$\rho_{air} = \frac{m}{V}$$

$$m = (1.29 \text{ kg/m}^3)(600 \text{ m}^3) = 774 \text{ kg}$$

Now calculate the change in temperature, using $Q = mc\Delta T$

$$\Delta T = \frac{Q}{mc} = \frac{(167,850 \text{ J})}{(774 \text{ kg})(733 \text{ J/kg})} = 0.3 \text{ C}°$$

132   **KAPLAN**

(d) The formula for linear expansion is:

$$\Delta L = \alpha L \Delta T$$
$$\Delta L = (12 \times 10^{-6}/\text{C}°)(4.5 \text{ cm})(10° \text{ C} - 30° \text{ C}) = -0.001 \text{ cm or } -0.1 \text{ mm}$$

The negative answer indicates a contraction of the rod due to reduction in its temperature.

5. Answers:

(a) Knowing that if the person floats 90% under water, then the density of the person is 90% the density of water, the person's density is $(0.9)(1,000 \text{ kg/m}^3)$, or 900 kg/m^3. Using density equals mass divided by volume:

$$V = \frac{m}{\rho} = \frac{55 \text{ kg}}{900 \text{ kg/m}^3} = 0.061 \text{ m}^3$$

(b) Specific gravity is equal to the density of the object divided by the density of water. We have already determined that the density of the person is 900 kg/m^3. Thus, the specific gravity of the person is $\dfrac{900}{1,000}$, or 0.9.

(c) When the swimmer is floating, the net force on him is zero. Thus, the buoyant force is equal to the swimmer's weight:

$$F_\text{B} = mg = (55 \text{ kg})(9.8 \text{ m/s}^2) = 539 \text{ N}$$

(d) It is helpful to remember that each 10 meter depth of fresh water is equivalent to about 1 atmosphere of pressure:

$$P = \rho g h = (10^3 \text{ kg/m}^3)(9.8 \text{ m/s}^2)(10 \text{ m}) = 98,000 \text{ N/m}^2 \text{ or } 98,000 \text{ Pa}$$
$$1 \text{ atm} = 1.01 \times 10^5 \text{ Pa}$$

(Actually, it takes about 10.3 m of water to equal one Pascal of pressure.)

Thus, 40 meters of water equals about 4 atm, so the total pressure on the swimmer is now 5 atm, including the atmosphere.

We can calculate this more exactly, if necessary:

$$P_T = P_o + \rho g h = 1.01 \times 10^5 \text{ Pa} + (10^3 \text{ kg/m}^3)(9.8 \text{ m/s}^2)(40 \text{ m}) = 4.93 \times 10^5 \text{ Pa}$$

(e) i. Since the increase in pressure will compress the body and decrease its volume, a decrease in volume should mean an increase in density of the body.

(e) ii. If the body increases in density, it will displace less water, even though it is now submerged. This means the buoyant force on the body will be less. (Incidentally, swimmers diving to great depths often experience this decrease in buoyancy and find it more difficult to swim back to the surface.)

AP Physics
CHAPTER 6

6. Answers:

(a) Since the difference in air pressure P_1 and P_2 will be insignificant for a 7.2 meter height difference, we consider only the difference in water pressure, using zero for the height in the basement:

$$\rho g h_2 - \rho g h_1 = (1{,}000 \text{ kg/m}^3)(9.8 \text{ m/s}^2)(7.2) = 70,560 \text{ or } 7.06 \times 10^5 \text{ Pa}$$

Using Bernoulli's Equation to compare the basement and second levels:

$$P_1 + \rho g h_1 + \frac{1}{2}\rho v_1^2 = P_2 + \rho g h_2 + \frac{1}{2}\rho v_2^2$$

$$(1{,}000 \text{ kg/m}^3)(9.8 \text{ m/s}^2)(0) + \frac{1}{2}(1{,}000 \text{ kg/m}^3)(12.5 \text{ m/s})^2 =$$

$$(1{,}000 \text{ kg/m}^3)(9.8 \text{ m/s}^2)(7.2 \text{ m}) + \frac{1}{2}(1{,}000 \text{ kg/m}^3)(v_2)^2$$

$$v_2 = 3.9 \text{ m/s}$$

(b) The volume rate of flow, in cubic meters per second, is the product of velocity of flow and cross-sectional area of pipe:

$$Q = Av = (\pi)(0.015 \text{ m})^2(12.5 \text{ m/s}) = 0.009 \text{ m}^3/\text{s}$$

(c) The volume rate of flow must remain the same throughout the system, so the volume rate of flow out of the faucet is 0.009 m³/s.

(d) Using the continuity equation to compare velocities as the pipe narrows at the faucet on the second floor:

$$A_1 v_1 = A_2 v_2$$
$$\pi(0.015)^2(3.9 \text{ m/s}) = \pi(0.0075)^2(v)$$
$$v = 15.6 \text{ m/s}$$

Electricity and Magnetism

CHAPTER 7

I. ELECTROSTATICS

Electric charge is the fundamental quantity that underlies all electrical phenomena. There are two types of charges: *positive* and *negative*. Like charges repel each other and unlike charges attract each other. The force between charges can be found by applying *Coulomb's law*. The *electric field* around a charge is the force per unit charge exerted on another charge in its vicinity. *Work* must be done to move a charge in an electric field, and the work is related to the *potential difference* between two points in an electric field.

Key Terms

charge: the fundamental quantity which underlies all electrical phenomena

charging by conduction: transfer of charge by actual contact between two objects

charging by induction: transfer of charge by bringing a charged object near a conductor, then grounding the conductor

conservation of charge: law that states that the total charge in a system must remain constant during any process

coulomb: the unit for electric charge

Coulomb's law: the electric force between two charges is proportional to the product of the charges and inversely proportional to the square of the distance between them

electric field: the space around a charge in which another charge will experience a force; electric field lines always point from positive charge to negative charge

KAPLAN 135

electric potential: the amount of work per unit charge to move a charge from a very distant point to another point in an electric field

electric potential difference: the difference in potential between two points in an electric field; also known as voltage

electron: the smallest negatively charged particle

electrostatics: the study of electric charge, field, and potential at rest

elementary charge: the smallest existing charge; the charge on one electron or one proton $(1.6 \times 10^{-19} \text{ C})$

grounding: the process of connecting a charged object to the earth or a large conductor to remove its excess charge

positively charged: having a deficiency of electrons

negatively charged: having an excess of electrons

neutral: having no net charge

test charge: the very small positive charge used to test the strength of an electric field

volt: the SI unit of potential or potential difference

Charge

Charge is the fundamental quantity that underlies all electrical phenomena. The symbol for charge is q, and the SI unit for charge is the *Coulomb* (C). The fundamental carrier of negative charge is the electron, with a charge of -1.6×10^{-19} C. The proton, found in the nucleus of any atom, carries exactly the same charge as the electron, but is positive. The neutron, also found in the nucleus of the atom, has no charge. When charge is transferred, only electrons move from one atom to another. Thus, the transfer of charge is really just the transfer of electrons. We say that an object with a surplus of electrons is negatively charged, and an object having a deficiency of electrons is positively charged. Charge is conserved during any process, and so any charge lost by one object must be gained by another object.

The Law of Charges

The law of charges states that like charges repel each other and unlike charges attract each other. This law is fundamental to understanding all electrical phenomena.

As an exercise, consider four charges, *A*, *B*, *C*, and *D*, which exist in a region of space. Charge *A* attracts *B*, but *B* repels *C*. Charge *C* repels *D*, and *D* is positively charged. What is the sign of charge *A*?

136 **KAPLAN**

Physics Review

ELECTRICITY AND MAGNETISM

If D is positive and it repels C, C must also be positive. Since C repels B, B must also be positive. A attracts B, so A must be negatively charged.

Charge is one of the four quantities in physics that is conserved during any process. Consider two charged spheres of equal size carrying a charge of +6 C and –4 C, respectively. The spheres are brought in contact with one another for a time sufficient to allow them to reach an equilibrium charge. They are then separated. We can find the final charge on each sphere by recognizing that charge will be transferred, but the total amount of charge is conserved.

The total charge on the two spheres is +6 C + –4 C = +2 C, and this is the magnitude of the equilibrium charge. When they are separated, they divide the charge evenly, each keeping a charge of +1 C.

Coulomb's Law

We know that two charges exert either an attractive or repulsive force on each other, but what is the nature of this force? It turns out that the force between any two charges follows the same basic form as Newton's law of universal gravitation, that is, the electric force is proportional to the magnitude of the charges and inversely proportional to the square of the distance between the charges. The equation for Coulomb's law is

$$F_E = \frac{Kq_1q_2}{r_2}$$

where F_E is the electric force, q_1 and q_2 are the charges, r is the distance between their centers, and K is a constant which happens to equal 9×10^9 N·m^2/C^2.

Sometimes the constant K is written as $K = \dfrac{1}{4\pi\varepsilon_o}$, where $\varepsilon_o = 8.85 \times 10^{-12}\ \dfrac{C^2}{N \cdot m^2}$.

As an example, consider two point charges q_1 and q_2 which are separated by a distance r, as shown above. The following choices refer to the electric force on these two charges:

(A) It is quadrupled
(B) It is doubled
(C) It remains the same
(D) It is halved
(E) It is quartered

What happens to the force on q_2 if the charge on q_1 is doubled?

KAPLAN 137

If q_1 is doubled, we have that $F = \dfrac{K(2q_1)(q_2)}{r^2} = 2F_E$. Thus the new force between the charges is doubled, or answer choice B.

What happens to the force on q_2 if the charge on q_1 is doubled and the distance between the charges is also doubled?

If q_1 is doubled and r is doubled, we have that $F = \dfrac{K(2q_1)(q_2)}{(2r)^2} = \dfrac{2}{4}F_E = \dfrac{1}{2}F_E$. Thus, the new force is half as much as the original force, or answer choice D.

Separation and Transfer of Charge in an Electroscope

An electroscope is a device which consists of a metal ball or plate connected to a metal rod with two thin metal leaves attached at the bottom. The rod and leaves are insulated so as not to pick up any extra charges from the air. Remember, even neutral objects have charges in them; they just have an equal number of positive and negative charges. We can use other charged objects to redistribute the charges in a neutral object without actually changing the amount of charge.

We can use an electroscope to study how charges in the ball, rod, and leaves are separated when we bring another charged object near the electroscope or touch the ball of the electroscope with the charged object. For example, your physics teacher may have charged a hard rubber rod negatively by rubbing it on a piece of fur. The rod becomes negatively charged because it strips electrons off the fur. If we bring the negatively charged rubber rod near the ball of the electroscope, the charges on the electroscope are separated. Use the diagram of the rod and electroscope below to draw how the positive and negative charges are distributed on the electroscope. Then check your answer below.

Did you draw the ball as positive and the leaves as negative and repelling each other? The free electrons in the ball are repelled by the nearby negatively charged rod, and therefore move down to the leaves of the electroscope. Since both leaves are then negative, they repel each other. We say that the ball of the electroscope is positively charged since it now has a lack of electrons. Your drawing should look like this:

But what if we touch the ball of the electroscope with the negatively charged rod? Draw the distribution of charges below:

The negative charges (electrons) flow into the electroscope, and we say that the electroscope is now negatively charged. This time, we didn't just separate charges in the metal; we actually transferred charge from one object to another by *conduction,* bringing the two objects in contact with each other. Your drawing should look like this:

Conductors, like metals, have electrons which are loosely bound to the outskirts of their atoms, and can therefore easily move from one atom to another. An *insulator,* like wood or glass, does not have many loosely bound electrons, and therefore cannot pass charge easily.

Electric Field

An electric field is the condition of space around a charge (or distribution of charges) in which another charge will feel a force. Electric field lines always point in the direction that a positive charge would feel a force. For example, if we take a charge Q to be the source of an electric field

E, and we bring a very small positive "test" charge *q* nearby to test the strength and direction of the electric field, then *q* will feel a force which is directed radially away from *Q*.

The electric field is given by the equation

E = **F***q*, where electric field **E** is measured in Newtons per coulomb, and **F** is the force acting on the charge *q* which is feeling the force in the electric field. Electric field is a *vector* which points in the same direction as the force acting on a positive charge in the electric field. The test charge *q* would feel a force radially outward anywhere around the source charge *Q*, so we would draw the electric field lines around the positive charge *Q* like this:

Remember, electric field lines in a region are always drawn in the direction that a *positive* charge would feel a force in that region. They can also represent the path a positive charge would follow in that region. We would draw the electric field lines around a negative charge, two positives, two negatives, and a positive and a negative charge as shown in the diagrams below.

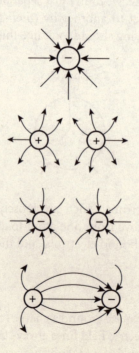

Remember, electrons (*negative* charges) are moved when charge is transferred, but electric field lines are drawn in the direction a *positive* charge would move.

The electric field due to a point charge Q at a distance r away from the center of the charge can also be written using Coulomb's law:

$$E = \frac{F}{q} = \frac{\left(\dfrac{KQq}{r^2}\right)}{q} = \frac{KQ}{r^2}$$

where K is the electric constant, Q is the source of the electric field, and q is the small charge which feels the force in the electric field due to Q.

Electric Potential

The *electric potential V* is defined in terms of the work we would have to do on a charge to move it against an electric field. For example, if we wanted to move a positive charge from point A to point B in the electric field shown below, we would have to do work on the charge, since the electric field would push against us.

We say that there is a potential difference ΔV between points A and B. The equation for potential difference between two points is:

$\Delta V = \dfrac{Work}{q}$, and is measured in *joules/coulomb*, or *volts*.

When we apply potential difference to circuits in a later section, we will often call it *voltage*. If we place the charge q at point B and let it go, it will "fall" toward point A. We say that positive charges naturally want to move from a point of *high potential* (B) to one of *low potential* (A), and we refer to the movement of the positive charges as *current*. We will return to voltage and current in the next section.

As we did for the electric field, we can write the electric potential due to a source charge Q at a distance r from the source charge:

$$V = \frac{KQ}{r}$$

Unlike electric field, which is a vector quantity, *electric potential* is a *scalar* quantity, that is, there is no direction or angle associated with potential, and potentials may be added without worrying about components. However, the potential due to a positive charge is positive, and the potential due to a negative charge is negative.

Remember, electric *field* is the *force* per unit charge and is a *vector*, and electric *potential* is the *work* per unit charge and is a *scalar*.

Electric Potential Energy

Recall that there is a gravitational potential energy that exists between two masses which is equal to the work that had to be done to bring the two masses from an infinite distance apart to a separation distance r. Similarly, there exists an electrical potential energy between two charges q_1 and q_2 which is equal to the work needed to bring the two charges from an infinite distance apart to a separation distance r:

$$U_E = \frac{Kq_1q_2}{r}$$

Since this expression is equal to the work done to bring the charges together, we can write $U_E = W = qV$, where q is the charge being moved in the electric field of the other charge.

Uniform Electric Field

We can create a uniform electric field in a region of space by taking two metal plates, setting them parallel to each other and separating them by a distance d, and placing a voltage V (like from a battery) across the plates so that one of the plates will be positive and the other negative.

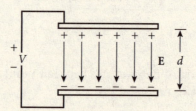

The positive charges on the top plate will line up uniformly with the negative charges on the bottom plate so that each positive charge lines up with a negative charge directly across from it. This arrangement of charges creates electric field lines which run from the positive charges to the negative charges and are uniformly spaced to produce a uniform (constant) electric field everywhere between the plates. Conducting plates which are connected this way are called a *capacitor*. Capacitors are used to store charge and electric field in a circuit which can be used at a later time. We will discuss capacitors further in a later section.

The electric field, voltage, and distance between the plates are related by the equation:

$$E = \frac{V}{d}$$

It follows from this equation that the unit for electric field is volts/meter, which is equivalent to newtons/coulomb.

Sample Problem:

Consider the electric field line diagrams below. Estimate what the magnitude of the charge on the right is in all cases. The charge on the left is +3 μC.

(a)

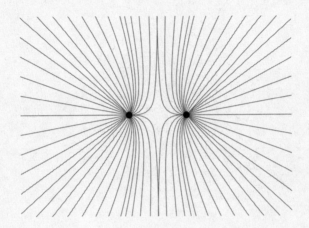

Answer:

+3 μC. The field lines repel each other, and thus the two charges must have the same sign. Since the picture is left/right symmetric the charges have the same magnitude.

(b)

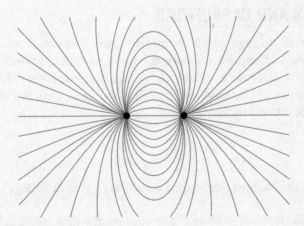

Answer:

−3μC. The field lines go from one charge to the other, and thus the two charges must have opposite signs. Since the picture is left/right symmetric the charges have the same magnitude.

(c)

Answer:
–1μC. More field lines originate on the left one than on the right. Therefore the magnitude of the left charge is bigger than that on the right. They must have opposite signs, because some field lines go from one to the other.

II. CONDUCTORS AND CAPACITORS

A *conductor* is a material through which charge can easily flow due to a large number of free electrons. Two equally and oppositely charged conductors, usually metal plates, which are near each other form a *capacitor* in which electrical energy and charge can be stored. The capacitance of a capacitor can be increased by filling the space between the conductors with an insulating material called a *dielectric*.

Key Terms

capacitance: ratio of the charged stored on a conductor per unit voltage

capacitor: electrical device used to store charge and energy in an electric field

conductor: a material through which heat or electric current can easily flow

dielectric: an electrically insulating material placed between the plates of a capacitor to increase its capacitance

dielectric constant: the ratio of the capacitance of a capacitor with a dielectric to its capacitance without a dielectric

farad: the unit for capacitance equal to one coulomb per volt

insulator: a material that is a poor conductor of heat or electric current due to a poor supply of free electrons

Charge on a Conductor

A conductor is a material, like a metal, through which charge is easily transferred. But if a charge is placed on a conductor, it immediately goes to the outside surface of the conductor. No charge remains on the inside because each charge would repel every other charge and they would get as far away from each other as possible. Once a free charge reaches the outside surface of a conductor, it has nowhere else to go, and thus remains on the outside surface until a way of escape is provided, such as grounding. Consider the metal sphere of radius R shown below. If a charge is placed anywhere in or on the metal, the charge will spread out symmetrically around the outside of the sphere. Let's sketch the electric field and electric potential as a function of distance r from the center of the sphere both inside and outside the sphere.

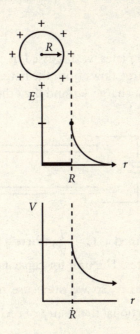

Note that the electric field is zero from $r = 0$ to $r = R$. This is because there is no free charge on the inside of the sphere, but all of it resides on the outside. At $r = R$, the electric field is $\frac{KQ}{R^2}$, and for $r \geq R$ the electric field decreases according to the inverse square law $\left(\frac{1}{r^2}\right)$.

The electric potential, however, is not zero inside the sphere. If we bring a small positive charge from infinity to the surface of the sphere at R, the potential increases since we are doing work against the electric field to move the charge toward the sphere. But once we go on the inside of the sphere, there is no electric field against which to do any work. Thus, from the surface of the sphere to its center, the potential does not change, and remains at $V = \dfrac{KQ}{R}$.

If we have two metal spheres of unequal size, we can transfer charge between them by connecting them with a wire:

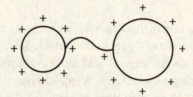

The charge will redistribute itself so that both spheres will be at the same *potential*, but not the same charge. If the two spheres are connected by a wire, they are essentially one surface which is at one potential. We call a surface which is at one potential an *equipotential surface*.

Capacitance

Let's revisit the conducting parallel plates we discussed in an earlier section. Recall that if we connect a battery to the plates, charge flows from the battery onto the plates, with one plate becoming positively charged with a charge $+q$, and the other plate negatively charged with a charge $-q$.

The capacitance of the plates is defined as $C = \dfrac{q}{V}$, where q is the charge on one of the plates, and V is the voltage across the plates. The unit for capacitance is the *coulomb/volt*, or *farad*. One farad is a very large capacitance, so we often use microfarads (μF), or 10^{-6} F. The capacitance of a capacitor is proportional to the area of each plate and inversely proportional to the distance between the plates. In symbols:

$$C \propto \dfrac{A}{d}$$

The constant of proportionality which makes the proportion above into an equation is ε_o, which is called the *permittivity of free space* and is equal to $8.85 \times 10^{-12} \dfrac{C^2}{N \cdot m^2}$. The permittivity constant gives us an indication of how well space holds an electric field. The equation for a parallel plate capacitor is:

$$C = \dfrac{\varepsilon_o A}{d}$$

Note that the capacitance of a capacitor ultimately only depends on its geometric dimensions, like area and distance between the plates. This is true for a capacitor of any shape. The purpose of a capacitor is to store charge and electric field in a circuit which can be used at a later time.

Let's say a capacitor has a capacitance C_0. What is the effect on the capacitance if (a) the area of each plate is doubled and (b) the distance between the plates is halved?

(a) If the area of each plate is doubled (assuming the plates have equal area), we can find the new capacitance by

$C \propto \dfrac{(2)A}{d} = 2C_0$. Twice the area gives twice the capacitance.

(b) If the distance between the plates is halved, that is, the plates are brought closer together, the new capacitance is

$C \propto \dfrac{A}{\frac{1}{2}d} = 2C_0$. Bringing the plates closer together increases the capacitance, in this case by a factor of two.

The electrical energy stored in a capacitor can be found by relating its charge q, voltage V, and capacitance C:

$$U_E = \frac{1}{2}CV^2 = \frac{q^2}{2C} = \frac{1}{2}qV$$

Capacitors in Parallel

There are times when we want to know the equivalent capacitance of two or more capacitors which are connected in a certain way. Consider two capacitors which are connected to a battery in the figure below:

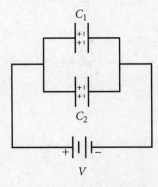

We say that the capacitors are connected in *parallel*. Note that the positive plates of the capacitors are connected to each other and the negative plates are connected to each other. Since the positive plates are also connected to the positive terminal of the battery, and the negative plates are connected to the negative terminal of the battery, the voltage across each capacitor is the same as the battery voltage. In other words, capacitors in parallel have the same voltage across them. However, the total charge will be divided proportionally among them:

$$q_{total} = q_1 + q_2$$

$$C_{total}V_{total} = C_1V_1 + C_2V_2$$

But the voltages are all the same:

$$C_{total} = C_1 + C_2 \text{ (parallel)}$$

Thus, to find the total capacitance in parallel, we simply add the capacitors.

Capacitors in Series

If we connect two capacitors in *series* as shown in the figure below, each capacitor will get the same charge, but will divide the voltage proportionally.

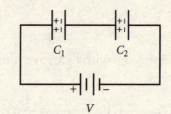

We can find the total (or equivalent) capacitance by summing the voltages:

$$V_{total} = V_1 + V_2$$

$$\frac{q_{total}}{C_{total}} = \frac{q_1}{C_1} + \frac{q_2}{C_1}$$

But, since all the charges are same,

$$\frac{1}{C_{total}} = \frac{1}{C_1} + \frac{1}{C_2} \text{ (series)}$$

For example, find the equivalent capacitance of the capacitors below.

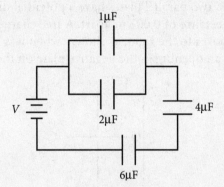

We can simply add the 1 μF and the 2 μF capacitors, since they are in parallel: 1 μF + 2 μF = 3 μF. But since the 3 μF, 4 μF, and 6 μF capacitors are in series, we must add their inverses to find their equivalent capacitance:

$$\frac{1}{C_{total}} = \frac{1}{3\mu F} + \frac{1}{4\mu F} + \frac{1}{6\mu F} \text{ implies that } C_{total} = \frac{4}{3}\mu F$$

Capacitor With a Dielectric

We can increase the capacitance of a parallel plate capacitor by inserting an insulating material, such as plastic or oil, between the plates of the capacitor:

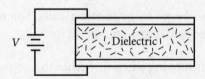

The insulating material is called a *dielectric*. Filling the space between the capacitor plates with a dielectric decreases the electric field between the plates, but increases the amount of charge that can be stored on the capacitor, and thus increases its capacitance. The amount by which the capacitance increases is called the dielectric constant, κ. For example if a particular dielectric has a constant $\kappa = 4$, the same-sized capacitor with this dielectric between its plates would have four times the capacitance as one without it.

AP Physics
CHAPTER 7

Sample Problem:

As shown below, two parallel plates have a potential difference of 12 volts and are held a distance of 0.001 m apart. A free charge held near the positive plate moves to the right, as shown, when it is released. It then moves through an opening in the negative plate on the right.

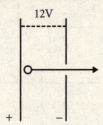

(a) Is the particle a proton or an electron?

Answer: Since the charge is repelled by the positive plate and attracted to the negative plate, it must be a proton.

(b) What is its energy (in Joules) as it leaves the right plate?

Answer: Calculate its potential energy at the positive plate:

$$U = qV = (1.6 \times 10^{-19} \text{ C})(12 \text{ V}) = 1.92 \times 10^{-18} \text{ J}$$

Note: This answer could also be given in electron volts, where 1 eV is the energy of one electron or one proton at a potential of 1 volt, or 12 eV.

(c) How fast is it moving when it reaches the negative plate?

Answer: The potential energy of the proton at the positive plate is converted to kinetic energy as it reaches the negative plate:

$$\Delta U = -\Delta K$$

$qV = \frac{1}{2}mv^2$, where m is the mass of a proton

$$v = (2qV/m)^{1/2} = \left[\frac{(2)(1.92 \times 10^{-18})}{(1.67 \times 10^{-27})}\right]^{\frac{1}{2}} = 4.80 \times 10^4 \text{ m/s}$$

(d) What are the magnitude and direction of the electric field between the plates?

Answer: Electric field strength can be calculated by knowing the potential difference and distance between the plates: $E = \frac{V}{d} = \frac{12 \text{ V}}{0.001 \text{ m}} = 12{,}000 \text{ V/m}$.

The direction of the field is from high potential to low potential, so the field is from left to right on the diagram.

III. ELECTRIC CIRCUITS

Conventional current is the flow of positive charges though a closed circuit. The current through a *resistance* and the *voltage* which produces it are related by *Ohm's law*. Resistors in a circuit may be connected in *series* or in *parallel*. If a *capacitor* is placed in a circuit with a resistor, the current in the circuit becomes time-dependent as the capacitor charges or discharges.

Key Terms

ammeter: device used to measure electrical current

ampere: unit of electrical current equal to one coulomb per second

battery: device that converts chemical energy into electrical energy, creating a potential difference (voltage)

capacitive time constant: the product of the resistance and the capacitance in a circuit; at a time equal to the product RC, the capacitor has reached 63% of its maximum charge

direct current: electric current whose flow of charges is in one direction only

electric circuit: a continuous closed path in which electric charges can flow

electric current: flow of charged particles; conventionally, the flow of positive charges

electron flow: the movement of electrons through a conductor; electron flow is equal and opposite to conventional current flow

emf: electromotive force; another name for voltage, particularly voltage induced in a conductor by electromagnetic induction

equivalent resistance: the single resistance that could replace the individual resistances in a circuit and produce the same result

kilowatt hour: amount of energy equal to 3.6×10^6 joules, usually used in electrical measurement

ohm: the SI unit for resistance equal to one volt per ampere

Ohm's law: the ratio of voltage to current in a circuit is a constant called resistance

parallel circuit: an electric circuit which has two or more paths for the current to follow, allowing each branch to function independently of the others

power: the rate at which work is done or energy is dissipated through a resistor

resistance: the ratio of the voltage across a device to the current running through it

AP Physics
CHAPTER 7

resistivity: the constant which relates the resistance of a resistor to its length and cross-sectional area

resistor: device designed to have a specific resistance

schematic diagram: a diagram using special symbols to represent a circuit

series circuit: an electric circuit in which devices are arranged so that charge flows through each equally

watt: the SI unit for power equal to one joule of energy per second

Current, Voltage, and Resistance

When we connect a battery, wires, and a light bulb in the circuit shown below, the bulb lights up. But what is actually happening in the circuit?

Recall from an earlier section that the battery has a potential difference, or voltage, across its ends. One end of the battery is positive, and the other end is negative. When we connect the wires and light bulb to the battery in a complete circuit, charge begins to flow from one end of the battery through the wires and the bulb to the other end of the battery, causing the bulb to light. We say that the movement of positive charge from the positive end of the battery through the circuit to the negative end of the battery is called *conventional current*, or simply *current*. Current is the amount of charge moving through a conductor per second, and the unit for current is the coulomb/second, or ampere. We use the symbol I for current.

Technically speaking, positive charges do not "flow" through a circuit, although it is common to speak of them as if they do. What is actually happening is when the battery is connected to the circuit above, the electrons on the negative side of the battery shift away from the negative end of the battery, bumping into electrons in the wire, creating a domino effect through the wire toward the positive end of the battery. This "wave" of bumping electrons heats the light bulb filament, which glows as a result, giving off light and heat. The AP Physics B exam generally uses the conventional current model of current flow.

As charge moves through the circuit, it encounters *resistance*, or opposition to the flow of current. Resistance is the electrical equivalent of friction. In our circuit above, the wires and the light bulb would be considered resistances, although usually the resistance of the wires is neglected. The resistance of a resistor is proportional to the length l of the resistor and inversely proportional to the cross-sectional area A of the resistor by the equation

$$R = \frac{\rho l}{A},$$

where the constant ρ is called the *resistivity* of the resistor and has units of ohm-meters ($\Omega \cdot m$). The resistivity of a material is a characteristic of the material and tends to vary with temperature. Sometimes we prefer to speak of the *conductivity* σ of a material rather than its resistivity. These are reciprocals of each other, so that $\sigma = \frac{1}{\rho}$.

In a circuit such as the one shown above, if a larger voltage is used, more charge moves through the wire, and thus more current flows through the circuit. In other words, the current is directly proportional to the voltage. This implies that the ratio of voltage to current is a constant. This constant is defined as the *resistance*, and is measured in *ohms* (Ω). An ohm is a volt/amp. The relationship between voltage, current, and resistance is called *Ohm's law*:

$$R = \frac{V}{I}$$

This relationship typically holds true for the purposes of the AP Physics B exam.

Schematic Diagram Symbols for Circuits

The table below summarizes the quantities discussed so far.

Quantity	Symbol	Unit	Schematic Symbol
Battery Voltage	V	Volts (V)	—⊣⊢—
Current	I	Amps (A)	→→
Resistance	R	Ohms (Ω)	—⋀⋀⋀—

Usually, when we draw circuit diagrams, we use the symbols in the table above. For example, the simple light bulb circuit would look like this:

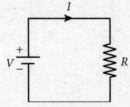

If we wanted to actually measure the current through the resistor and the voltage across the resistor, we would connect an *ammeter* (to measure current) in series with the resistor, and a *voltmeter* (to measure voltage) in parallel with the resistor, as shown below:

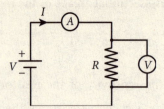

We place the ammeter in *series* with the resistor so that the same current will pass through the ammeter and the resistor, and we place the voltmeter in *parallel* with the resistor so that the voltage will be the same across the voltmeter and the resistor. Ammeters have a low resistance so as not to add to the total resistance of the circuit (and thus decrease the current in the circuit), and voltmeters have a high resistance so that current will not want to flow through them and bypass the resistor.

Power and Joule's Law of Heating

In an earlier section we defined *power* as the rate at which work is done, or the rate at which energy is transferred. When current flows through a resistor, heat is produced, and the amount of heat produced in joules per second is equal to the power in the resistor. The heating in the resistor is called *joule heating*, and the power dissipated in the resistor follows *Joule's law of heating*. The equation that relates power to the current, voltage, and resistance in a circuit is

$$P = IV = I^2R = \frac{V^2}{R}$$

The unit for power is the *joule/second*, or *watt*.

For example, a simple circuit consists of a 12-volt battery and a 6 Ω resistor. Let's draw a schematic diagram of the circuit, and include an ammeter which measures the current through the resistor, and a voltmeter which measures the voltage across the resistor. Indicate the reading on the ammeter and voltmeter, and find the power dissipated in the resistor.

The schematic diagram would look like this:

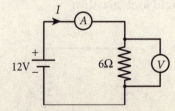

Note that the ammeter is placed in series with the resistor, and the voltmeter is placed in parallel with the resistor.

The ammeter reads the current, which we can calculate using Ohm's law:

$$I = \frac{V}{R} = \frac{12\text{ V}}{6\text{ Ω}} = 2A$$

The voltmeter will read 12 V, since the potential difference across the resistor must be equal to the potential difference across the battery. As we will see later, if there were more than one resistor in the circuit, they each would not necessarily get 12 volts.

The power can be found by:

$$P = IV = (2 \text{ A})(12 \text{ V}) = 24 \text{ watts}$$

Circuits with More Than One Resistor

When two or more resistors are placed in a circuit, there are basically three ways to connect them: in series, in parallel, or in a combination of series and parallel.

Series Circuit

Two or more resistors of any value placed in a circuit in such a way that the same current passes through each of them is called a *series circuit*. Consider the series circuit below which includes a voltage source ε (which stands for *emf*, an older term for voltage) and three resistors: R_1, R_2, and R_3.

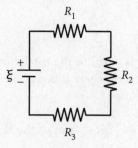

The rules for dealing with *series* circuits are as follows:

1. The total resistance in a series circuit is the sum of the individual resistances:

$$R_{total} = R_1 + R_2 + R_3$$

2. The total current in the circuit is

$$I_{total} = \frac{V_{total}}{R_{total}}$$

This current must pass through each of the resistors, so each resistor also gets I_{total}, that is, $I_{total} = I_1 = I_2 = I_3$.

3. The voltage divides proportionally among the resistances according to Ohm's law:

$$V_1 = I_1 R_1;\ V_2 = I_2 R_2;\ V_3 = I_3 R_3;$$

As an example, consider three resistors of 2 Ω, 4 Ω, and 6 Ω connected in series with a 12 volt battery. Let's draw a schematic diagram of the circuit, which includes an ammeter to measure the current through the 2 Ω resistor, and a voltmeter to measure the voltage across the 6 Ω resistor, and indicate the reading on the ammeter and voltmeter.

The schematic diagram for the circuit looks like this:

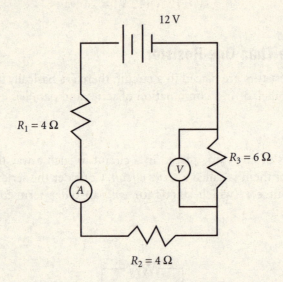

The current through each resistor is equal to the total current in the circuit, so the ammeter will read the total current regardless of where it is placed, as long as it is placed in series with the resistances.

The total resistance in the circuit, R_T, is the sum of the resistors in parallel, or 14 ohms.

The total current in the circuit is calculated using Ohm's Law:

$$I_{meter} = I_T = \frac{V_T}{R_T} = \frac{12 \text{ v}}{14 \Omega} = 0.86 \text{ A}$$

The voltmeter will read the voltage across the 6 Ω resistor, which is **not** 12 volts. The 12 volts provided by the battery are divided proportionally among the resistances. The voltage across the 6 Ω resistor is:

$$V_6 = I_6 R_6 = (0.86 \text{ A})(6 \text{ Ω}) = 5.2 \text{ V}$$

Parallel Circuits

Two or more resistors of any value placed in a circuit in such a way that each resistor has the same potential difference across it is called a *parallel circuit*. Consider the parallel circuit below which includes a voltage source ε and three resistors R_1, R_2, and R_3.

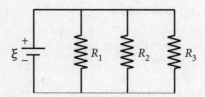

The rules for dealing with *parallel* circuits are as follows:

1. The total resistance in a parallel circuit is given by the equation:

$$\frac{1}{R_{total}} = \frac{1}{R_1} + \frac{1}{R_2} + \frac{1}{R_3}$$

2. The voltage across each resistance is the same:

$$V_{total} = V_1 = V_2 = V_3$$

3. The current divides in an inverse proportion to the resistance:

$$I_1 = \frac{V_1}{R_1}; I_2 = \frac{V_2}{R_2}; I_3 = \frac{V_3}{R_3}$$

where V_1, V_2, and V_3 are equal to each other.

For example, the resistors of resistance 2 Ω, 4 Ω, and 8 Ω are connected in parallel to a battery of voltage 12 V.

Let's draw a schematic diagram of the circuit, which includes an ammeter to measure only the current through the 2 Ω resistor and a voltmeter to measure the voltage across the 8 Ω resistor. Then find the total resistance in the circuit, along with the readings on the ammeter and voltmeter.

The schematic diagram for the circuit looks like this:

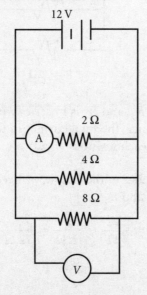

The total resistance in the circuit is found by:

$$\frac{1}{R_{total}} = \frac{1}{2} + \frac{1}{4} + \frac{1}{8} = \frac{7}{8}$$

Notice that this fraction is not the total resistance, but $\frac{1}{R_{total}}$. Thus, the total resistance in this circuit must be $\frac{8}{7}\Omega$.

Since the ammeter is placed in series only with the 2 Ω resistance, it will measure the current passing only through the 2 Ω resistance. Recognizing that the voltage is the same (12 V) across each resistance, we have:

$$I_2 = \frac{V_2}{R_2} = \frac{12\ V}{2\ \Omega} = 6A$$

Each resistance is connected across the 12 V battery, so the voltage across the 8 Ω resistance is 12 V.

Combination Series and Parallel Circuits

Consider the combination circuit below:

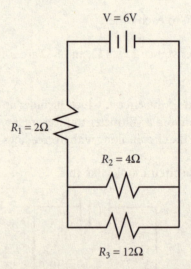

We see that R_3 is in parallel with R_2, and R_1 is in series with the parallel combination of R_3 and R_2. Let's find the total resistance, the total current in the circuit, the voltage across each resistor, and the current through each resistor.

Before we can find the total resistance of the circuit, we need to find the equivalent resistance of the parallel combination of R_3 and R_2:

$$\frac{1}{R_{32}} = \frac{1}{R_2} + \frac{1}{R_3} = \frac{1}{4\ \Omega} + \frac{1}{12\ \Omega} = \frac{3}{12\ \Omega} + \frac{1}{12\ \Omega} = \frac{4}{12\ \Omega}$$

This implies that the combination of R_3 and R_2 has an equivalent resistance of

$$\frac{12}{4}\Omega = 3\ \Omega.$$

Then the total resistance of the circuit is $R_{total} = R_{32} + R_1 = 3\ \Omega + 2\ \Omega = 5\ \Omega$.

The total current in the circuit is $I_{total} = \dfrac{V_{total}}{R_{total}} = \dfrac{6\ V}{5\ \Omega} = 1.2\ A$.

The voltage provided by the battery is divided proportionally among the parallel combination of R_3 and R_2 (with R_3 and R_2 having the same voltage across them), and R_1. Since R_1 has the total current passing through it, we can calculate the voltage across R_1:

$$V_1 = I_1 R_1 = (1.2\ A)(2\ \Omega) = 2.4\ V$$

This implies that the voltage across R_3 and R_2 is the remainder of the 6 V provided by the battery. Thus, the voltage across R_3 and R_2 is 6 V − 2.4 V = 3.6 V.

The current through R_1 is the total current in the circuit, 1.2 A. Since we know the voltage and resistance of the other resistances, we can use Ohm's Law to find the current through each.

$$I_3 = \frac{V_3}{R_3} = \frac{3.6\ V}{12\ \Omega} = 0.3 A$$

$$I_2 = \frac{V_2}{R_2} = \frac{3.6\ V}{4\ \Omega} = 0.9 A$$

Notice that I_3 and I_2 add up to the total current in the circuit, 1.2 A.

Kirchhoff's Rules for a Multi-loop Circuit

Consider the multi-loop circuit shown below:

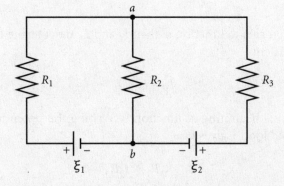

In this circuit, we have three resistors and two batteries, one on either side of junction b. We can't say that all the resistors are in series with each other, or that they are parallel, because of the placement of the batteries. If we want to find the current in and voltage across each resistor, we must use *Kirchhoff's rules*.

Kirchhoff's Rules for Multi-loop Circuits:

1. The total current entering a junction (like *a* or *b* in the figure above) must also leave that junction. This is sometimes called the *junction rule*, and is a statement of *conservation of charge*.

2. The sum of the potential rises and drops (voltages) around a loop of a circuit must be zero. This is sometimes called the *loop rule*, and is a statement of *conservation of energy*.

    a. If we pass a battery from negative to positive, we say that there is a rise in potential, $+\varepsilon$.

    b. If we pass a battery from positive to negative, we say that there is a drop in potential, $-\varepsilon$.

    c. If we pass a resistor against the direction of our arbitrarily chosen current, we say there is a rise in potential across the resistor, $+IR$.

    d. If we pass a resistor in the direction of our arbitrarily chosen current, we say there is a drop in potential across the resistor, $-IR$.

Let's apply these rules to the circuit shown above. First, let's choose arbitrary directions for each of the currents through the resistors. If we happen to choose the wrong direction for a particular current, the value of the current will simply come out negative.

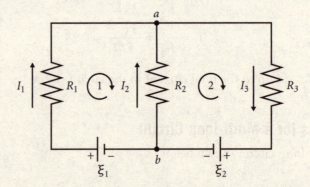

If we apply the junction rule to junction *a*, then I_1 and I_2 are entering the junction, and I_3 is exiting the junction. Then:

$$I_1 + I_2 = I_3$$

Let's apply the loop rule beginning at junction *b*. Writing the potential rises and drops by going clockwise around loop 1 gives us:

$$+\varepsilon_1 - I_1 R_1 + I_2 R_2 = 0$$

Note that we encountered a rise in potential $(+\varepsilon)$ as we passed the battery, a drop $(-I_1 R_1)$ across the first resistor, and a rise $(+I_2 R_2)$ across the second resistor.

Similarly, if we write the potential rises and drops around loop 2, going clockwise beginning at b, we get:

$$-I_2R_2 - I_3R_3 - \xi_2 = 0$$

By substitution, we can solve the three equations above for the values of the three currents if we know the values of the resistors and the emfs of the batteries.

Resistance-Capacitance Circuits

Recall from an earlier section that a *capacitor* is made up of two conductors which are charged oppositely and are separated by an empty space or a dielectric. The unit for a capacitor is the *farad* (F), and is equal to *one coulomb per volt*. A *resistance-capacitance* (RC) circuit is simply a circuit containing a battery, a resistor, and a capacitor in series with one another. An RC circuit can store charge and release it at a later time. The capacitive time constant τ_c is equal to the product of R and C (which has units of time), and gives an indication of how long it takes for a capacitor to charge or discharge. Typically, when connected to a battery, a capacitor is fully charged in a time of approximately 5RC.

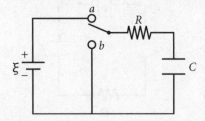

When the switch is moved to position a, current begins its flow from the battery, and the capacitor begins to fill up with charge. Initially, the current is $\frac{\varepsilon}{R}$ by Ohm's law, but it decreases as time goes by until the capacitor is full of charge and will not allow any more charge to flow out of the battery.

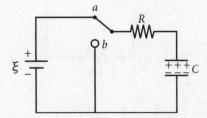

This leads us to a couple of rules we can follow when dealing with capacitors in an RC circuit:
1. An empty capacitor does not resist the flow of current, and thus acts like a wire.
2. A capacitor which is full of charge will not allow current to flow, and thus acts like a broken wire.

If we move the switch to position *b*, the battery is taken out of the circuit and the capacitor begins to drain its charge through the resistor, creating a current in the opposite direction to the current flowing when the battery was connected.

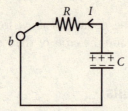

Eventually, the current will die out because of the heat energy lost through the resistor.

For example, a 6-volt battery is connected in series to a 4 Ω resistor and an initially uncharged capacitor. Determine the current in the circuit immediately after the switch is closed and again a long time after the switch has been closed.

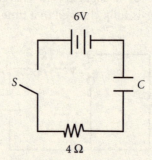

Immediately after the circuit is connected, the capacitor is still empty and thus acts like a wire. We can redraw the circuit like this:

The current, then, is $I = \dfrac{V}{R} = \dfrac{6\,V}{4\,\Omega} = 1.5A$

Then the current begins to decrease as the capacitor fills up with charge, and after a long time the capacitor is full of charge and the current stops flowing completely. Thus, $I = 0$ a long time later.

ELECTRICITY AND MAGNETISM

Sample Problem:

1. In the circuit below, a circuit breaker stops the flow of current (by opening the circuit) when the current exceeds 15 A. This protects the house from possible fire from the heat produced in the wiring. The circuit is for a bedroom that will have a television (12 Ω) turned on first, then a hair dryer (15 Ω), and finally a vacuum cleaner (10 Ω). When each new appliance is turned on, the others remain on. We can model the house power as a 120 V battery.

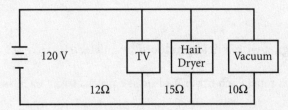

   (a) Which appliance will cause the breaker to trip when it is turned on?

Answer:

Because it is a parallel circuit, each appliance receives the full 120 V of voltage, so the television alone draws $\frac{120 \text{ V}}{12 \text{ Ω}} = 10$ A of current, which is within the safe limit.

Next, the hair dryer draws $\frac{120 \text{ V}}{15 \text{ Ω}} = 8$ A. This current, in addition to the 10 A from the television, produces a total current of 18 A, which exceeds the rating of the circuit breaker.

Thus, the hair dryer trips the breaker.

   (b) To the nearest 5A, what kind of current rating (the maximum current that can flow through the circuit breaker without opening the circuit) must a circuit breaker have in order to allow all three appliances to operate simultaneously?

Answer:

When you add the vacuum, you add an additional $\frac{120 \text{ V}}{10} = 12$ A of current to the circuit. This makes a grand total of 10 A + 8 A + 12 A = 30 A.

Thus, a 30 A breaker would allow all the appliances to run simultaneously.

   (c) What is the power rating of the hair dryer?

Answer: $P = VI = (120\text{V})(8 \text{ A}) = 960$ W

IV. MAGNETOSTATICS

Magnetostatics is the study of magnetic fields which do not change with time. A *magnetic field* is the space around a magnet in which another magnet will experience a force. Magnetic poles can be *north* or *south*, and like poles repel each other and unlike poles attract. Fundamentally, magnetism is caused by moving charges, such as a current in a wire. Thus, a moving charge or current-carrying wire produces a magnetic field, and will experience a force if placed in an external magnetic field.

Key Terms

electromagnet: a magnet with a field produced by an electric current

law of poles: like poles repel each other and unlike poles attract each other

lodestone: a naturally occurring magnetic rock made principally of iron

magnetic domain: cluster of magnetically aligned atoms

magnetic field: the space around a magnet in which another magnet or moving charge will experience a force

magnetostatics: the study of magnetic fields which do not change with time

right-hand rules: used to find the magnetic field around a current-carrying wire or the force acting on a wire or charge in a magnetic field

Magnetism

Since the earth itself is a magnet, a lodestone made of iron ore buried in the ground would become magnetized in the earth's magnetic field. If we suspend a magnet from a string, one end seeks magnetic north and the other seeks magnetic south. Thus, the two ends of a magnet are commonly called the north pole and the south pole. Magnetic north and south are near geographic south and north, respectively. (Remember that the north pole of a magnet, which points "north", would only be attracted by a south magnetic pole.) Any magnet has a magnetic field around it. A *magnetic field* **B** is defined as the space around a magnet in which another magnet will feel a force, and is measured in *Teslas* (T). Magnetic field lines have no beginning and no end, but generally are drawn from the north pole of a magnet to its south pole. The magnetic fields most often appearing on the AP Physics exam are those of a bar magnet, and a horseshoe magnet, each shown below.

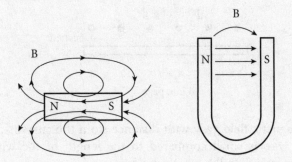

A material becomes magnetic when it is placed in a strong external magnetic field, and the clusters of atoms with similar magnetic orientations, called *domains*, become aligned with the external magnetic field. Obviously, some materials are more easily magnetized than others, but the reasons are typically not covered on the AP Physics B exam.

Magnetic Field Produced by a Current-Carrying Wire

Prior to 1820, it was generally believed that there was no physical connection between electricity and magnetism. But in that year, Oersted discovered that a current-carrying wire is magnetic, or as we would say in today's language, it creates a magnetic field around itself. Fundamentally, magnetic fields are produced by moving charges. This is why all atoms are tiny magnets, since the electrons around the nucleus of the atom are moving charges and are therefore magnetic. The magnetic field due to a current-carrying wire circulates around the wire in a direction given by what we will refer to as the *first right-hand rule*:

First Right-hand Rule: Place your thumb in the direction of the current *I*, and your fingers will curl around in the direction of the magnetic field produced by that current.

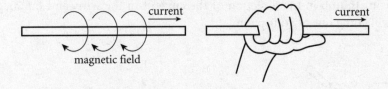

In determining the direction of a magnetic field due to the flow of electrons in a wire, we would use the left hand instead of the right hand.

Another way of denoting any vector which points into the page or out of the page, in this case the magnetic field vector, is to use an *X* to represent a vector pointing into the page, and a dot (•) to represent a vector pointing out of the page. [Note: If it is difficult for you to remember, think of an arrow. When it is coming toward you, the point is visible first, and when it is moving away, the "tail feathers" are visible.] Thus, the magnetic field around the current-carrying wire above can be drawn as:

AP Physics
CHAPTER 7

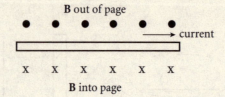

But how does the magnetic field vary with distance from the current-carrying wire? If the distance r from the wire is small compared to the length of the wire, we can find the magnitude of the magnetic field B by the equation

$$B = \frac{\mu_o I}{2\pi r},$$

where μ_o is called the *permeability constant* and is equal to $4\pi \times 10^{-7} \frac{T \cdot m}{A}$. The value 4π in this constant is often used for reasons of geometry. The magnetic field around a current-carrying wire is proportional to $\frac{1}{r}$, while electric field around a point charge is proportional to $\frac{1}{r^2}$.

Force on a Current-Carrying Wire in an External Magnetic Field

Since a current-carrying wire creates a magnetic field around itself according to the first right-hand rule, every current-carrying wire is a magnet. Thus, if we place a current-carrying wire in an external magnetic field, it will experience a force. The direction of the force acting on the wire is given by what we will call the *second right-hand rule*.

Second Right-hand Rule: Place your fingers in the direction of the magnetic field (north to south), and your thumb in the direction of the current in the wire, and the magnetic force on the wire will come out of your palm.

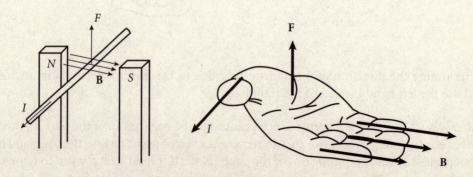

Again, you would use your left hand to find the direction of the magnetic force if you were given electron flow instead of conventional current.

The equation for finding the force on a current-carrying wire in a magnetic field is

$$F = ILB \sin \theta$$

where I is the current in the wire, L is the length of wire which is in the magnetic field, B is the magnetic field, and θ is the angle between the length of wire and the magnetic field. If the angle is 90°, the equation becomes simply $F = ILB$.

For example, consider a wire carrying a 20 A current and having a length $L = 0.10$ m between the poles of a magnet at an angle of 45°, as shown. The magnetic field is uniform and has a value of 0.5 T. What is the magnitude and direction of the magnetic force acting on the wire? ($\sin 45° = \cos 45° = 0.7$)

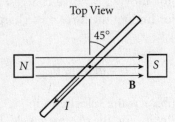

The magnitude of the force on the wire is found by:

$$F = ILB \sin \theta = (20\ A)(0.10\ m)(0.5\ T) \sin 45°$$
$$F = 0.7\ N$$

The direction of the force can be found by the second right-hand rule. Place your fingers in the direction of the magnetic field, and your thumb in the direction of the length (and current) which is perpendicular to the magnetic field, and we see that the force is out of the page.

Note that the length must have a component which is perpendicular to the magnetic field, or there will be no magnetic force on the wire. In other words, if the wire is placed parallel to the magnetic field, $\sin 0° = 0$, and the force will also be zero.

Remember, use your *right* hand for current or moving positive charges, and your *left* hand for electron flow or moving negative charges.

If we bend a wire into a square loop and pass a current through the loop while it is in a magnetic field, each side of the loop will experience a force:

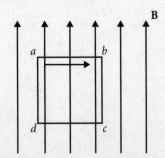

In this case, the magnetic field is uniform and toward the top of the page. The current is clockwise in the loop. By the second right-hand rule, side *ab* will experience a force, out of the page, and side *cd* will experience a force into the page. The current in sides *bc* and *da* is parallel to the magnetic field, so there is no magnetic force acting on them. The result is a torque on the loop, causing it to rotate in the magnetic field. This is the basic principle behind ammeters, voltmeters, and the electric motor.

Force on a Charged Particle in a Magnetic Field

Since a moving charge creates a magnetic field around itself, it will also feel a force when it moves through a magnetic field. The direction of the force acting on such a charge is given by the second right-hand rule, with the thumb pointing in the direction of the velocity of the charge. We use our right hand for moving positive charges, and our left hand for moving negative charges. The equation for finding the force on a charge moving through a magnetic field is:

$$F = qvB \sin \theta$$

where q is the charge in Coulombs, v is the velocity in m/s, B is the magnetic field in Teslas, and θ is the angle between the velocity and the magnetic field. If the angle is 90°, the equation becomes $F = qvB$.

Consider a proton entering a magnetic field **B** which is directed into the page. The proton has a charge $+q$ and a velocity **v** which is directed to the right, and enters the magnetic field perpendicularly.

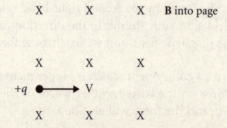

As the proton enters the magnetic field, it will initially feel a force which is directed upward on the page, as we see from using the second right-hand rule. The path of the proton will curve toward the top of the page in a circular path, with the magnetic force becoming a centripetal force, changing the direction of the velocity to form the circular path. The radius of this circle can be found by setting the magnetic force equal to the centripetal force:

magnetic force = centripetal force

$$qvB = \frac{mv^2}{r}$$

$$r = \frac{mv}{qB}$$

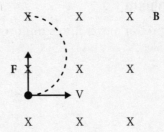

But what if we wanted to cause the charge to follow a straight-line path through the magnetic field, that is, pass through the field undeflected? If we place the magnetic field between two charged parallel plates, we can orient the electric field to apply a force on the moving charge that is equal and opposite to the magnetic force. In this case, the electric force on the charge would need to be directed downward on the page to counter the magnetic force upward on the page. Thus, the top plate would be positive, and the bottom plate would be negative, so that the electric field between the plates would be directed downward on the page, as shown below:

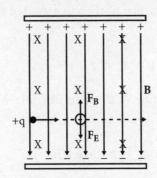

$$\mathbf{F}_{net} = \mathbf{F}_E + \mathbf{F}_B$$

$$\mathbf{F}_{net} = q\mathbf{E} + q\mathbf{v}\mathbf{B}$$

The net force acting on the moving charge is the sum of the electric and magnetic forces, and is called the *Lorentz force*:

$$qE = qvB$$

$$E = vB$$

This expression relates the speed of the charge and the electric and magnetic fields for a charge moving undeflected through the fields.

AP Physics
CHAPTER 7

Sample Problem:

A proton moves into a uniform magnetic field at a 45 degree diagonal, as shown.

```
X   X   X   X   X   X
X   X   X   X   X   X     B = 1.2T
X   X   X   X   X   X
X   X   X   X   X   X
X   X   X   X   X   X
X   X   X   X   X   X
```

$v = 2.2 \times 10^3$ m/s

(a) What is the magnitude of the force exerted on the proton as it enters the field?

Answer:

$$F = \left| qvB \right| = (1.6 \times 10^{-19} \text{ C})(2.2 \times 10^3 \text{ m/s})(1.2 \text{ T}) = 4.2 \times 10^{-16} \text{ N}$$

(b) What is the direction of the magnetic force?

Answer:

$$\boldsymbol{F} = q\, \mathbf{v} \times \boldsymbol{B}$$

Using the right hand rule, the velocity is toward the top right corner of the page, the field is directed perpendicular to the page and into the page, and the force is directed diagonally in the plane of the page toward the upper left corner of the page.

(c) The proton will take a curved path as it enters the field. Calculate the radius of the circular path it begins to take.

Answer: The magnetic force will provide the centripetal force for the proton's circular path.

$$qvB = \frac{mv^2}{R}$$

$$R = \frac{mv}{qB}$$

$$= \frac{(1.67 \times 10^{-27} \text{ kg})(2.2 \times 10^3 \text{ m/s})}{(1.6 \times 10^{-19} \text{ C})(1.2 \text{ T})}$$

$$= 1.9 \times 10^{-5} \text{ m}$$

170 **KAPLAN**

Physics Review

ELECTRICITY AND MAGNETISM

V. ELECTROMAGNETISM

Since a current produces a magnetic field, we may assume that a magnetic field can produce a current. *Electromagnetic induction* is the process by which an *emf* (or *voltage*) is produced in a wire by a changing *magnetic flux*. Magnetic flux is the product of the magnetic field and the area through which the magnetic field lines pass. Electromagnetic induction is the principle behind the *electric generator* and the *transformer*. The direction of the induced emf or current is governed by *Lenz's law*.

Key Terms

alternating current: electric current that rapidly reverses its direction

electric generator: a device that uses electromagnetic induction to convert mechanical energy into electrical energy

electromagnetic induction: inducing a voltage in a conductor by changing the magnetic field around the conductor

induced emf: the voltage produced by electromagnetic induction

Faraday's Law of Induction: law which states that a voltage can be induced in a conductor by changing the magnetic field around the conductor

Lenz's Law: the induced emf or current in a wire produces a magnetic flux which opposes the change in flux that produced it by electromagnetic induction

magnetic flux: the product of the magnetic field and the area through which the magnetic field lines pass.

transformer: device which uses electromagnetic induction to transfer energy from one circuit to another

Electromagnetic Induction

For several years after Oersted discovered that a current flowing through a wire produced a magnetic field. Several scientists, including Faraday in England and Henry in New York, tried to verify whether or not the opposite were true: Can a magnetic field produce a current? Faraday and Henry independently discovered that an electric current could be produced by moving a magnet through a coil of wire, or, equivalently, by moving a wire through a magnetic field. Generating a current this way is called *electromagnetic induction*.

KAPLAN 171

AP Physics
CHAPTER 7

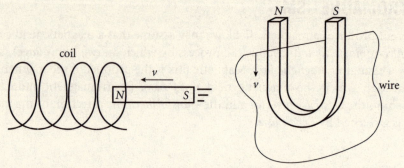

Moving a wire into a magnetic field.

The amount of voltage and current produced in a coil of wire depends on how quickly the magnetic field lines are crossed by the wire. For example, if the magnet is moved through the coil slowly, hardly any current is produced in the coil. If the magnet is moved through the coil quickly, a larger amount of current is produced in the coil. Also, a greater number of coils will produce a greater induced voltage and current.

All of the following can increase the amount of voltage induced in a coil of wire:

(A) move the magnet faster through the coil
(B) move a stronger magnet through the coils of wire
(C) move the magnet through more coils of wire
(D) move more coils of wire around a magnet
(E) move more magnets simultaneously through a coil of wire

The principle of electromagnetic induction is the basis for a generator. A generator converts mechanical energy into electrical energy. Place a loop of wire on an axle in a magnetic field, as shown below. As the loop is rotated, the wire crosses magnetic field lines and generates a current in the loop. That current can be used to light a light bulb, or to power a city. All of our electrical power is generated in a similar way.

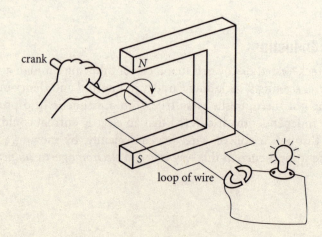

But how can we tell which direction an emf or current is induced in a wire? Consider a wire moving downward through the magnetic field shown below:

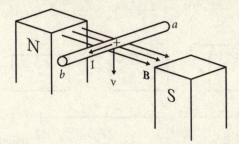

Imagine a little positive charge inside the wire. As the wire moves downward, the positive charge moves downward as well. If we apply the second right-hand rule to determine the direction of the force on the positive charge in the wire, we see that the force on the charge is away from point *a* and toward point *b*. As a result of electromagnetic induction, the charge will move toward point *b*, and thus an emf will be induced in the wire with point *b* becoming the positive end of the wire and point *a* becoming the negative end. If the wire is part of a complete circuit, a current will flow from point *a* to point *b*.

Consider a rectangular loop of wire of height *l* and width *w* that sits in a region of magnetic field of length $3l$. The magnetic field is directed into the page, as shown below:

The loop is in the plane of the page so that its area is perpendicular to the magnetic field lines. The *magnetic flux* Φ through the loop is the scalar product of the magnetic field **B** and the area **A** through which it fluxes:

$$\Phi = \mathbf{B} \cdot \mathbf{A}$$

As the loop sits at rest in the magnetic field, the magnetic flux through it is constant, and is equal to $BA = Blw$. If the loop were tilted at an angle θ relative to the magnetic field, the flux would be

$$\Phi = BA \sin \theta,$$

where θ is the angle between the field direction and plane of the loop.

But if the loop is moved through the magnetic field with a constant speed v, the flux through the loop does not remain constant, but rather changes with time:

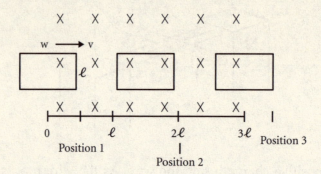

Faraday's Law of Induction states that an emf ξ will be induced in a loop of wire through which the flux is changing:

$$\xi = \frac{-\Delta \Phi}{\Delta t}$$

Note that the induced emf in the wire is not caused by the magnetic flux or field, but by the *change in flux* through the loop.

It may occur to you that we could use Faraday's Law of Induction to create an emf or current, and the induced current would produce a new magnetic field and flux which could produce more current, and so on. But *Lenz's Law* tells us that this is not possible. Lenz's Law states that the induced emf or current in a wire produces a change in flux which opposes the change in flux that produced it. The new change in flux works against the old change in flux. This is essentially a statement of conservation of energy.

Let's find the magnitude and direction of the emf ξ and current I induced in the loop of resistance R. As the loop enters the region of magnetic field it begins enclosing magnetic field lines, as seen in Position 1 in the figure above. Let the width of the loop which is in the magnetic field be x. The emf induced in the loop as the loop is entering the magnetic field is

$$\xi = \frac{-\Delta \Phi}{\Delta t} = \frac{-\Delta (BA)}{\Delta t}$$

where A is the area of magnetic field which is enclosed by the loop of wire. In Position 1, this area would be lx. So, the equation for ε becomes

$$\xi = \frac{-B[\Delta(lx)]}{\Delta t} = -Bl\left(\frac{\Delta x}{\Delta t}\right) = -Blv$$

The negative sign simply indicates that the induced emf opposes the change in flux that produced it (Lenz's Law). The current can be found by Ohm's Law:

$$I = \frac{\xi}{R} = \frac{-B\ell v}{R}$$

The direction of the induced current in the loop can be found by applying the second right-hand rule as we did in the previous example, or by applying Lenz's law. As the loop enters the magnetic field, the flux is increasing inward. Thus the induced current will produce a flux which is increasing outward to oppose the original change in flux which produced it. The right side of the wire would need to have a current flowing upward to create an outward flux through the inside of the loop. The induced current would therefore flow in a counterclockwise direction around the loop.

When the loop is completely in the region of the magnetic field (Position 2), no current is induced in the loop since there is no *change* in flux through the loop. As the loop exits the field (Position 3), the flux through the loop is decreasing inward, so the induced current will produce a flux which opposes this decrease in flux, producing more flux inward. Thus, current will flow clockwise around the loop as the loop exits the field.

Let's plot graphs of flux Φ through the loop and induced emf ξ in the loop as functions of the width, x, of the wire in the field.

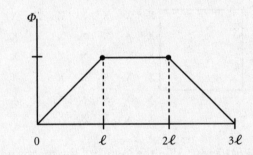

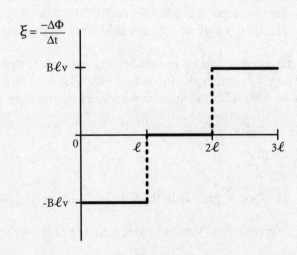

AP Physics
CHAPTER 7

The flux rises linearly from zero to *l*, is constant while the loop is completely in the field, then falls linearly as the loop exits the field. The induced ε is constant while the flux is changing, and zero while the flux is constant. The plot of the magnitude of the induced ε as a function of time is the *slope* of the Φ vs. *t* graph.

In reality, the loop would slow down slightly as it enters the field, since the induced current would create a magnetic force on the leading edge of the wire which opposes the direction of motion of the loop. The magnitude of this force is

$$F = ILB$$

where *I* is the induced current in the wire. By symmetry, the magnetic force on the wire would point in the direction of motion as the loop exits the field.

Sample Problem:

The square wire loop with sides of length *L* and resistance *R* is being pushed at velocity *v* through a magnetic field of strength *B*. Express all answers in terms of the given variables and fundamental constants.

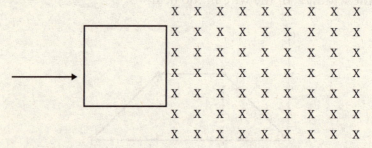

(a) Calculate EMF induced in the loop and the magnitude and direction of current induced in the loop in each of the following situations:
   i. The loop is halfway into the field and still moving to the right.

Answer: By Lenz's Law, current will be induced in the loop such that a magnetic field is produced out of the page to counter the increasing magnetic field into the page inside the loop as it moves into the field. Thus, a counterclockwise current will be induced in the loop.

The magnitude of the EMF is: $\xi = BLv$

The current is: $I = \dfrac{\xi}{R} = \dfrac{BLv}{R}$

   ii. The loop is entirely in the field and still moving to the right.

Answer: When the loop is entirely in the field, there is no net EMF on the loop, so no current flows.

ELECTRICITY AND MAGNETISM

Physics Review

iii.   The loop is halfway out of the field and still moving to the right

Answer: By Lenz's Law, current will be induced in the loop such that a magnetic field is produced into the page to counter the decreasing magnetic field into the page inside the loop as it moves out of the field. Thus, a clockwise current will be induced in the loop.

The magnitude of the EMF is: $\xi = BLv$

The current is: $I = \dfrac{\xi}{R} = \dfrac{BLv}{R}$

iv.   The loop is halfway back into the field and now moving to the left.

Answer: By Lenz's Law, current will be induced in the loop such that a magnetic field is produced out of the page to counter the increasing magnetic field into the page inside the loop as it moves into the field. Thus, a counterclockwise current will be induced in the loop.

The magnitude of the EMF is: $\xi = BLv$

The current is: $I = \dfrac{\xi}{R} = \dfrac{BLv}{R}$

v.   The loop is halfway back into the field but stopped.

Answer: When the loop stops moving, no EMF is induced, and no current flows in the loop

What will happen if the loop is now positioned entirely in the field and connected to a power supply so that a clockwise current flows through it?

Answer: If we use the right hand rule on the positive charge flow (direction of current) on each side of the wire, there is a magnetic force outward on each side of the loop. Thus, the loop attempts to expand in the field but does not move.

KAPLAN   177

AP Physics
CHAPTER 7

Important Equations:

Coulomb's law (electrostatic force):
$$F = \frac{kq_1q_2}{r^2} = \frac{\left(\frac{1}{4\pi\varepsilon_o}\right)q_1q_2}{r^2}$$

Electric field:
$$E = \frac{F}{q} = \frac{\left(\frac{1}{4\pi\varepsilon_o}\right)q_1q_2}{r}$$

Electric potential:
$$V = \frac{\left(\frac{1}{4\pi\varepsilon_o}\right)\Sigma_i q_i}{r_i}$$

Electric flux:
$$\Phi_E = \boldsymbol{E} \cdot \boldsymbol{S} = ES \cos\theta$$

Electric potential energy:
$$U_E = qV$$

Electric current:
$$I = \frac{\Delta q}{\Delta t}$$

Electric field of a point charge:
$$E = \frac{kq}{r} = \frac{\left(\frac{1}{4\pi\varepsilon_o}\right)q}{r}$$

Electric field between capacitor plates:
$$E_{ave} = -\frac{V}{d}$$

Capacitance:
$$C = \frac{q}{V}$$

Energy stored in a capacitor:
$$U = \left(\frac{1}{2}\right)CV^2 = \frac{1}{2}qV$$

Force on a moving charge in a magnetic field:
$$F = qvB \sin\theta = q\boldsymbol{v}\ \boldsymbol{X}\ \boldsymbol{B}$$

Force on a wire in a magnetic field:
$$F = I\,\boldsymbol{L}\,\boldsymbol{X}\,\boldsymbol{B} = I\,L\,B \sin\theta$$

Resistance of a wire:
$$R = \frac{\rho L}{A}$$

Ohm's law:
$$V = IR$$

Electrical power:
$$P = VI$$

Capacitors in series:
$$\frac{1}{C_s} = \Sigma_i \frac{1}{C_i}$$

ELECTRICITY AND MAGNETISM

Capacitors in parallel: $C_p = \Sigma_i\, C_i$

Resistors in series: $R_s = \Sigma_i\, R_i$

Resistors in parallel: $R_p = \Sigma_i\, R_i$

Magnetic field around a wire: $B = \dfrac{(\mu_0 I)}{(2\pi r)}$

Magnetic field in a solenoid: $B = \mu_0 n I$

Torque on a loop of wire in a magnetic field: $\tau = IAB \sin\theta$

Magnetic flux: $\Phi_B = \boldsymbol{B} \cdot \boldsymbol{A} = BA \cos\theta$

Induced potential difference (emf)
due to a changing magnetic field: $\varepsilon = -N\dfrac{\Delta\phi}{\Delta t}$

Motional emf: $\varepsilon = BLv$

AP Physics
CHAPTER 7

PRACTICE PROBLEMS

1. Use the picture of the two charges of +1 µC and +3 µC as shown below.
   (a) Draw the electric field lines of the charges.

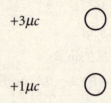

+3µc

+1µc

   (b) What will change in the picture if the charges are both negative?
   (c) Calculate the magnitude of the force between the two charges in part *a* if they are placed a distance of 2 cm apart.
   (d) What is the magnitude of the force between the two charges in part *b* if they are placed a distance of 2 cm apart?
   (e) What is the electric field due to the two charges in *a* at a point midway between them?
   (f) What is the electric field due to the two charges in *b* at a point midway between them?

2. A constant electric field of magnitude $E = 129.1$ V/m points in the positive *x*-direction.
   (a) How much work (in J) does it take to move a charge $Q = 3$ µC from $x_1 = -10$ m to $x_2 = 73$ m? *(Hint: The charge should move at a constant speed.)*

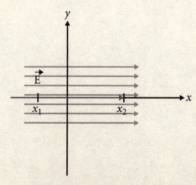

   (b) If the charge is released at the 73 meter mark, in what direction will it move? Justify your answer.
   (c) If, very close to the point the charge is released, the field ends, what will happen to the charge when the field ends?

ELECTRICITY AND MAGNETISM

3. A metal sphere of radius $R = 10$ cm carries a total charge $Q = 0.4$ μC.
   (a) What is the magnitude of the electric field just outside the sphere, and in which direction is it pointing?
   (b) What is the magnitude and direction of the electric field just inside the sphere?
   (c) A second sphere of the same radius carries a total charge $Q = +0.6$ μC. If it is brought near the first sphere but does not touch it, what happens to the charge on the first sphere
      i.   as the second sphere is brought close?
      ii. as it moves away?
   (d) If the second sphere is brought in so that it touches the first sphere, what happens to the charge on the first sphere?
   (e) Consider the situation again. Now as the second charged sphere is brought in to a position near the first charged sphere but is not touching it, the first sphere is grounded. Then the grounding line is disconnected and the second sphere is removed. What happens to the charge on the first sphere?

4. Four charges are arranged on the corners of a square with the length of each side equal to one meter. A fifth charge is located in the middle of the square.

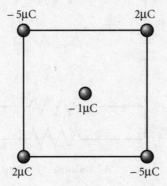

   (a) What is the absolute potential at the center of the square due to the four charges arranged at the corners?
   (b) What is the potential energy of the $-1\mu C$ charge as it is positioned at the center of the square?
   (c) What is the electric force on the charge in the center of the square due to the four charges at the corners?

5. In the circuit below, the switch is suddenly closed.
   (a) For time = 0, find the current and the potential difference across the capacitor.
   (b) After several time constants have passed, find the current, the potential difference across the capacitor, and the total charge stored on the capacitor.

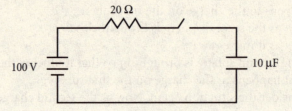

6. (a) In the given circuit, find the current flowing through the resistor marked A.

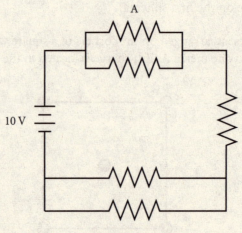

All Resistors = 10 ohms

   (b) All the resistors are replaced with capacitors of value 10 microfarads. Find the equivalent capacitance and the charge stored in capacitor A.

7. Consider this arrangement of charges:

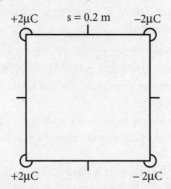

(a) Find the magnitude and direction of the electric field at the center of the square.
(b) What force (magnitude and direction) would a charge of $-4\ \mu C$ experience if it was placed at the center of the square?
(c) How would the electric field at the center of the square change if you removed the two negative charges on the right?
(d) Sketch the electric field lines for the two remaining charges and show how your drawing supports your answer to part c.

8. Consider the following arrangement of potentials and a positive test charge:

(a) To which location will a test charge of $+3\ \mu C$ travel?
(b) What will be its kinetic energy when it arrives?
(c) Find the capacitance of a capacitor that would store the same amount of energy as you found in part b, using the same potential difference.
(d) Suppose the $+3\ \mu C$ test charge is replaced with a test charge of $-3\ \mu C$. To which location will it travel?
(e) If the new test charge has a mass of 27.3×10^{-21} kg, what will be its speed when it reaches its destination?

9. A flash attachment for a professional camera stores energy in a capacitor. When a picture is taken, all of the charge is converted to energy, and the capacitor is fully discharged.

   (a) Assume the battery charging the capacitor is a 300 V battery. When the light flashes, it produces 5,000 W of light for a time of 0.005 s. Find the total energy produced by the flash.
   (b) Find the capacitance.
   (c) If the battery were replaced with a battery with a potential difference of 120 V, how much light power could the flash attachment produce?

10. A loop of area 4 cm^2 has its plane parallel to the field lines of the magnetic field $B = 0.6$ T, as shown in the figure. The loop is pulled in the opposite direction of the field with a constant velocity of $v = 6$ m/s. What is the induced voltage?

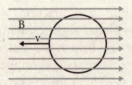

11. The drawing below shows a circular loop of wire (radius 0.15 m) connected to a power supply. The top loop is directly over the bottom loop, which is not connected to anything.

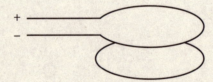

   (a) When the switch is first turned on, in what direction will the current first flow in the bottom loop?
   (b) The bottom coil receives a constant magnetic field of 0.8 T from the top coil and starts spinning perpendicular to its original plane at a speed of two revolutions per second. Find the emf induced in that coil as it rotates through one-fourth of a revolution.
   (c) If the bottom loop begins rotating away from the viewer (the near part of the loop begins to move down), in which direction is current first induced in that loop?

12. In the drawing below, an electron is released and allowed to accelerate through a potential difference between two charged plates as shown. It then travels horizontally through Parallel Plate Capacitor B until it reaches the magnetic field on the right, at which point if follows one of two paths.

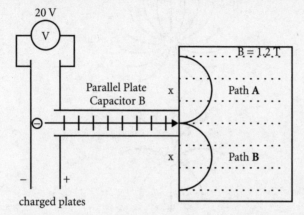

(a) With what speed does the electron leave the opening in the charged plates and enter parallel plate capacitor B?
(b) Determine the direction of the electric field in capacitor B and the polarity of the second capacitor that will produce the desired effect. *(Hint: Take into account the effects of gravity.)*
(c) Tell whether the electron will follow path A or path B as it passes through the magnetic field.
(d) Find the value of x (the diameter of the circle).

Solutions to Practice Problems

1. (a) Answers:

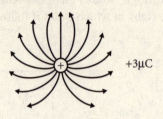

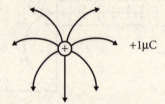

(b) Everything stays the same, except that now the arrows point in the opposite direction.

(c) $F = \dfrac{kq_1q_2}{r^2} = \dfrac{\left(\dfrac{9 \times 10^9 \text{ Nm}^2}{\text{C}^2}\right)(1 \times 10^{-6} \text{ C})(3 \times 10^{-6} \text{ C})}{(.02 \text{ m})^2} = 67.5 \text{ N}$

(d) If the charges are both negative, the magnitude of the force is the same.

(e) For the two positive charges, the electric field at the midway point due to the upper charge:

$$E = \dfrac{kq_1}{r^2} = \dfrac{\left(\dfrac{9 \times 10^9 \text{ Nm}^2}{\text{C}^2}\right)(3 \times 10^{-6} \text{ C})}{(0.01 \text{ m})^2} = 2.7 \times 10^8 \text{ N/C}$$

toward the bottom of the page.

The electric field at the midway point due to the lower charge:

$$E = \dfrac{kq_1}{r^2} = \dfrac{\left(\dfrac{9 \times 10^9 \text{ Nm}^2}{\text{C}^2}\right)(1 \times 10^{-6} \text{ C})}{(0.01 \text{ m})^2} = 9.0 \times 10^7 \text{ N/C}$$

toward the top of the page.

The **net electric field** is the sum of the two vector fields, which is 1.8×10^8 N/C toward the bottom of the page.

(f) For the two negative charges, the electric field at the midway point due to the upper charge:

$$E = \frac{kq_1}{r^2} = \frac{\left(\frac{9 \times 10^9 \text{ Nm}^2}{\text{C}^2}\right)(3 \times 10^{-6} \text{ C})}{(0.01 \text{ m})^2} = 2.7 \times 10^8 \text{ N/C}$$

toward the top of the page.

The electric field at the midway point due to the lower charge:

$$E = \frac{kq_1}{r^2} = \frac{\left(\frac{9 \times 10^9 \text{ Nm}^2}{\text{C}^2}\right)(1 \times 10^{-6} \text{ C})}{(0.01 \text{ m})^2} = 9.0 \times 10^7 \text{ N/C}$$

toward the bottom of the page.

The **net electric field** is the sum of the two vector fields, which is 1.8×10^8 N/C

toward the top of the page.

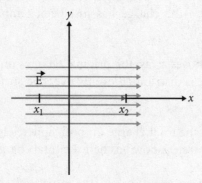

2.
Answers:
(a) In this case, the work being done in moving the charge from x_1 to x_2 is to stop the particle from accelerating. The work is equal to:

$$\begin{aligned}
W &= QE(x_2 - x_1) \\
&= (3 \times 10^{-6} \text{ C})(129.1 \text{ V/m})(73 \text{ m} - -10 \text{ m}) \\
&= (3 \times 10^{-6})(129.1)(83) \\
&= -0.032 \text{ J}
\end{aligned}$$

(b) Since the field is to the right and the charge is positive, the charge has a force on it to the right, so the charge will accelerate to the right when released at point x_2.

(c) The particle will no longer have a force on it, so it will continue to move to the right at the same speed it had when the field ended.

AP Physics

CHAPTER 7

3. Answers:

(a) Recall that the electric field outside the sphere is the same as that of a point charge at the center of the sphere and is therefore equal to:

$$E = \frac{kq}{r^2}$$

$$= \frac{\left(\dfrac{9 \times 10^9 \text{ Nm}^2}{C^2}\right)(0.4 \times 10^{-6} \text{ C})}{(10 \times 10^{-2} \text{ m})^2}$$

$$= 360{,}000 \text{ N/C}$$

The field is pointing outward in radial direction.

(b) Inside the metallic sphere, the excess charges repel each other, rearranging themselves on the surface of the sphere so that the net electric field inside the sphere is zero.

(c) i. As the second sphere, which also has a positive charge, is brought near the first, a negative charge is induced on the side of the first sphere facing the second sphere. The second sphere repels positive charges and attracts negative charges. However, the total charge on the first sphere remains the same–the charge has just been temporarily redistributed by the process of induction.

(c) ii. As the second sphere moves away, the original charges on the first sphere redistribute again so that the total charge is distributed evenly over the surface of the sphere, with total charge remaining the same.

(d) When the spheres touch, the total charge on both spheres is evenly distributed between the two spheres, so that each sphere now has half the total charge, or 0.5 microcoulombs

(e) If the first sphere had been grounded in the situation where the second sphere was brought near without actually touching, the positive charges on the first sphere would be repelled by the second sphere, causing the positive charges on the first sphere to move to the ground. If the grounding wire is cut while the second sphere is still near, the first sphere will be left with a less positive charge or net negative charge. Another way of thinking about this is to consider instead that negative charges move onto the first sphere from the ground, attracted to the positive charge on the nearby second sphere. When the grounding wire is cut, the first sphere is left with less positive charge—the "new" negative charges having neutralized the initial positive charge on the sphere.

4. Answers:

(a) The electric potential at the center is not a vector; it is the scalar sum of the potentials due to the four charges. Since the length of each side of the square is one meter, some simple trigonometry yields the distance, r, from each charge to the center as $\sqrt{2}$, or $r = 1.4$ m. The total potential is the sum of the potentials due to each of the charges at the corners:

188 **KAPLAN**

Physics Review

ELECTRICITY AND MAGNETISM

$$\Sigma V = \left(\frac{k}{r}\right)(\Sigma q) = \frac{\left(\dfrac{9 \times 10^9 \text{ Nm}^2}{C^2}\right)(-6 \times 10^{-6} \text{ C})}{1.4 \text{ m}} = -3.9 \times 10^4 \text{ V}$$

(b) The easiest method of finding the potential energy of the charge in the center is to first find the potential, as we did in part (a), then use the formula $U = qV$:

$$U = (1 \times 10^{-6} \text{ C})(3.9 \times 10^4 \text{ V}) = 0.039 \text{ J}$$

[Note: We use only the magnitude of the charge, since energy is a scalar quantity.]

(c) It is important when dealing with vector quantities to first note any symmetries, or situations where components of the vectors added will cancel. In this case, the forces due to the two $+2\mu C$ charges will cancel each other. The forces on the center charge due to the two $-5\mu C$ charges are also equal and opposite, so they cancel.

Thus, there is no net force on the center charge.

5. Answers:

(a) At time = 0, no charge is yet stored on the capacitor, so there is no voltage across it. Because there is no voltage across the capacitor, all 100 V must drop across the resistor. Therefore, from $V = IR$, we find that the current, $I = \dfrac{V}{R} = \dfrac{100 \text{ V}}{20 \text{ }\Omega} = 5 \text{ A}$.

(b) After several time constants, the capacitor will be fully charged. This means that no more current will flow across the resistor. Thus, $I = 0 \text{ A}$. Since no current flows, there can't be any voltage drop across the resistor. This means that all 100 V drops across the capacitor. Thus, $V = 100 \text{ V}$. From the expression for capacitance, $q = CV = (10 \times 10^{-6} \text{ F})(100 \text{ V}) = 0.001 \text{ C}$.

6. Answers:
(a) Since all resistors are 10 ohms, the top parallel portion reduces to a total resistance of 5 ohms. Likewise, the bottom parallel portion also reduces to 5 ohms. Combining the resulting series resistors (5 Ω, 10 Ω, and 5 Ω) we get a total circuit resistance of 20 Ω. Using Ohm's law ($V = IR$) this gives us a total current of 0.5 A in the circuit. Since the resistor marked A is in a parallel circuit and is equivalent to the other resistor in the circuit, each receives the same current. So, the current must "split" down the parallel link. Therefore, the current in resistor "A" = 0.25 A.

(b) Replace all resistors with capacitors, each with a capacitance of 10 μF. Find the charge stored on the capacitor marked A.

KAPLAN 189

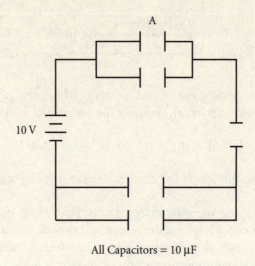

All Capacitors = 10 μF

Combining the top capacitors (parallel) gives us an equivalent capacitance of 20 μF. Likewise, the bottom parallel capacitors also have an equivalent capacitance of 20 μF. When you add up three capacitors in series (20 μF, 10 μF, and 20 μF), you get a total capacitance of 5 μF. Using $q = CV$, we find that the total charge stored in the circuit is 50 μC. This means that each equivalent part of the circuit stores 50 μC, but capacitor A is part of a parallel circuit in which each parallel capacitor stores an equal charge. Therefore, the capacitor marked A stores 25 μC.

7. Answers:

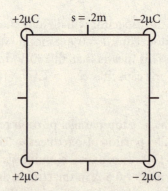

(a) The E field at the center due to each of the corner charges will have the same magnitude:

$$E = \frac{F}{q} = \frac{kq}{r^2} = \frac{(9 \times 10^9)(2 \times 10^{-6})}{(0.1 \times 2^{\frac{1}{2}})^2} = 900{,}000 \text{ N/C}$$

Since the fields are vectors and the fields due to the upper left and lower right charges each point toward the lower right, their magnitudes can be added. The same goes for the upper right and lower left charges: the net field is twice the original magnitude and points toward the upper right.

Since these two vectors are perpendicular to each other, they can be added using the Pythagorean Theorem:

$E_{total} = [(2 \times 900{,}000)^2 + (2 \times 900{,}000)^2]^{1/2} = 2{,}545{,}584$ N/C $= 2.55 \times 10^6$ N/C, to the right.

(b) The force felt is:

$$F = Eq = (2.55 \times 10^6 \text{ N/C})(4 \times 10^{-6} \text{ C}) = 10.2 \text{ N}$$

The direction is opposite the direction of the field (since the charge is negative) and, thus, points to the left.

(c) If you removed the two charges on the right, the direction of the electric field would still point to the right; it would just be weaker, since only two charges are producing it instead of four. The electric field strength would be half as large.

(d) This drawing shows the two positive charges, with the location of the center of the square marked with the arrow. At that point, all field lines point to the right.

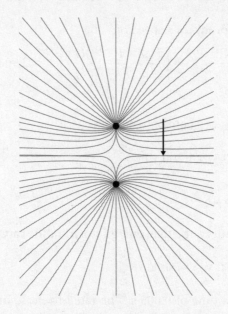

8. Answers:

(a) Positive charges seek the lowest potential, so the test charge will travel to the -2 V location.

(b) The initial potential of the test charge is $+5$ V, so it travels through a potential difference of 7 V. Voltage is defined as energy/charge.

$7V = \dfrac{E}{(3 \times 10^{-6} \text{ C})}$. So:

$$E = (7V)(3 \times 10^{-6} \text{ C}) = 2.1 \times 10^{-5} \text{ J}$$

AP Physics
CHAPTER 7

Thus, the kinetic energy of the test charge is 2.1×10^{-5} J.

(c) The energy stored in a capacitor is:

$$E = \left(\frac{1}{2}\right)CV^2$$

We know that $E = 2.1 \times 10^{-5}$ J, and that $V = 7$ V.

Substituting and solving for C gives you:

$$C = 0.857 \ \mu F$$

(d) A negative charge will seek the point of highest potential, so it will travel to the 9 V location.

(e) The energy of the new test charge is:

$$E = qV = (-3 \times 10^{-6} \ C)(4 \ V) = 12 \times 10^{-6} \ J = 1.2 \times 10^{-5} \ J$$

This energy, when kinetic, produces a speed of:

$$V = \left[\frac{(2)(KE)}{m}\right]^{1/2} = \left[\frac{(1.2 \times 10^{-5} \ J)(2)}{(27.3 \times 10^{-21})}\right]^{1/2} = 2.96 \times 10^7 \ m/s$$

9. Answers:

(a) Energy = Power × Time = $(5{,}000 \ W)(0.005 \ s) = 25$ J

(b) Energy = $\left(\frac{1}{2}\right)CV^2 = (0.5)(C)(300 \ V)^2$

Thus, $C = \dfrac{(25 \ J)(2)}{(300 \ V)^2} = 556 \ \mu F$

(c) When we replace the battery, the total energy stored in the capacitor changes:

$E = (1/2)(555 \times 10^{-6} \ F)(120 \ V)^2 = 3.996$ J

This energy is released in a time of 0.005 s. The rate of release of energy is power: energy divided by time.

$$P = \frac{Energy}{Time} = \frac{3.996 \ J}{0.005 \ s} = 800 \ W$$

10. Answer:

No calculation is needed. The area normal vector of the loop is perpendicular to the magnetic field vector at all times. So $\cos \theta = \cos (90°) = 0$. The result: $V = 0$.

192   **KAPLAN**

Physics Review
ELECTRICITY AND MAGNETISM

11. Answers:

(a) When the current begins to flow in the top loop, the magnetic flux through the loop increases in the downward direction. Therefore, the induced current in the bottom loop will produce flux that opposes that change. That means the flux generated in the lower loop will point upward, and the current will be flowing counter-clockwise.

(b) Induced emf is produced by a changing flux. The time for $\frac{1}{4}$ of a revolution to occur is $\frac{1}{4}$ of the 0.5 seconds required for each revolution, or 0.125 s.

The magnetic field isn't changing, but the area through which it is passing does change. So:

$$e = -\frac{\Delta \Phi}{\Delta t} = \frac{(B \cdot \Delta A)}{\Delta t} = \frac{(0.8 \text{ T})(0.0707 \text{ m}^2)}{0.125 \text{ s}} = 0.45 \text{ V}$$

(c) The flux through the loop is decreasing in the downward direction as the rotation begins. To counteract that change, the induced current will produce flux that will increase in that downward direction. Thus, the direction of the induced current is clockwise.

12. Answers:

(a) The electron has energy equal to $qV = (1.6 \times 10^{-19} \text{ C})(20 \text{ V}) = 3.2 \times 10^{-18}$ J. This energy is then converted to kinetic energy:

$$\left(\frac{1}{2}\right)mv^2 = 3.2 \times 10^{18} \text{ J}$$

Solving for v, we find $v = 2.65 \times 10^6$ m/s.

(b) Since the electron would normally be accelerated toward the ground by gravity, there must be an upward force on it to keep it moving in a straight line. An upward force on an electron would be produced by an electric field that points downward (since electric fields point in the direction a positive charge will be accelerated). Thus, the top of the capacitor must be positively charged, and the bottom must be negatively charged.

(c) An electron moves opposite to the direction of conventional current flow. So, using the right-hand rule, and knowing that the magnetic field points out of the page, you determine that the initial force on the electron is upward on the page. The electron will travel path A.

(d) X is the diameter of the circle along which the electron travels. By setting centripetal force equal to magnetic force, we know that the radius is:

$$r = \frac{mv}{qB} = \frac{(9 \times 10^{-31} \text{ kg})(2.65 \times 10^6 \text{ m/s})}{(1.6 \times 10^{-19} \text{ C})(1.2 \text{ T})} = 1.26 \times 10^{-5} \text{ m}$$

But X is simply $2r = (2)(1.26 \times 10^{-5}) = 2.52 \times 10^{-5}$ m.

KAPLAN 193

Waves and Optics

CHAPTER 8

I. WAVE MOTION

A *wave* is a disturbance which causes a transfer of energy. *Mechanical waves* need a *medium* in which to travel, but *electromagnetic waves* do not. Waves can be *transverse* or *longitudinal*, depending on the direction of the vibration of the wave. *Sound* is a longitudinal wave. A wave is characterized by its *frequency, period, wavelength, amplitude,* and *speed*. Waves can be *reflected, refracted, or diffracted,* and two waves in the same medium will *interfere*.

Key Terms

amplitude: maximum displacement from equilibrium position; the distance from the midpoint of a wave to its crest or trough.

beat: variations in the loudness of sounds due to the slight difference in frequency of interfering waves

closed pipe resonator: a pipe closed at one end with a sound source at the other, causing the sound to resonate

constructive interference: addition of two or more waves which are in phase, resulting in a wave of increased amplitude

crest of a wave: the highest point on a wave

decibel: the unit for the loudness of a sound; one-tenth of a Bel

destructive interference: addition of two or more waves which are out of phase, resulting in a wave of decreased amplitude

diffraction: the spreading of a wave beyond the edge of a barrier or through an opening

frequency: the number of vibrations of a wave per unit of time

KAPLAN 195

hertz: the unit for frequency equal to one cycle or vibration per second

in phase: term applied to two or more waves whose crests and troughs arrive at a place at the same time in such a way as to produce constructive interference

interference of waves: displacements of two or more waves in the same medium at the same time producing either larger or smaller waves

law of reflection: the angle of incidence of an incoming wave is equal to the angle of reflection measured from a line normal (perpendicular) to the surface

longitudinal wave: wave in which the vibration of the medium is parallel to the direction of motion of the wave

loudness: the quality of a sound wave which is measured by its amplitude

mechanical wave: a wave which uses a material medium through which to transfer energy

node: the point of no displacement in a standing wave

out of phase: term applied to two or more waves for which the crest of one wave arrives at a point at the same time as the trough of a second wave arrives, producing destructive interference

period: the time for one complete cycle or revolution

periodic motion: motion that repeats itself at regular intervals of time

pitch: the perceived sound characteristic equivalent to frequency

principle of superposition: the displacement due to two or more interfering waves is equal to the sum of the displacements of the individual waves

refraction: the change in speed, wavelength, and direction of a wave due to a change in medium

standing wave: wave with stationary nodes produced by two identical waves traveling in opposite directions in the same medium at the same time

transverse wave: a wave in which the vibration is perpendicular to the velocity of the wave

trough of a wave: the low point of wave motion

wavelength: the distance between successive identical parts of a wave

General Wave Properties

A *mechanical wave* is a traveling disturbance in a medium which transfers energy from one place to another. A *medium* is the substance through which a wave moves, such as water for a water wave, or air for a sound wave. An *electromagnetic wave* is a vibration of an electric and magnetic field which travels through space at an extremely high speed, and does not need a medium through which to travel. Light, radio waves, and microwaves are all examples of electromagnetic waves. We will return to electromagnetic waves later.

There are two types of mechanical waves. *Transverse waves* vibrate in a direction which is perpendicular to the direction of motion of the wave. For example if you hold the end of a spring and vibrate your hand up and down, you create a transverse wave in the spring.

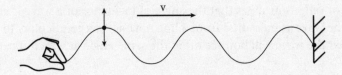

If you gather the spring up into a bunch and let it go, you create a *longitudinal wave*, in which the spring vibrates in a direction which is parallel to the direction of motion of the wave.

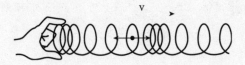

Sound is a common example of a longitudinal wave, since the air through which a sound wave moves is repeatedly compressed and expanded.

A wave has a *period*, the time it takes for a wave to vibrate once; a *frequency*, the number of waves that pass a given point per second; and an *amplitude*, the maximum displacement of a wave, or its height. A wave also has length. The length of one complete vibration of a wave is called the *wavelength*, and is denoted by the Greek letter lambda, λ. The figure below illustrates these quantities.

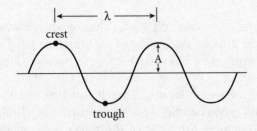

The *crest* is the highest point on the wave and the *trough* is the lowest point on the wave. The wavelength can be measured from one crest to the next crest, or from one trough to the next trough.

The speed of a wave can be found by the equation $v = f\lambda$, where v is the speed, f is the frequency, and λ is the wavelength. Since frequency is the reciprocal of period, we can also write this equation as $v = \dfrac{\lambda}{T}$. The speed of all types of waves can be found using this equation.

Reflection, Refraction, Diffraction, and Interference of Mechanical Waves

Let's discuss four things you can do with a wave. *Reflection* is the bouncing of a wave off of a barrier. The law of reflection states that the angle of incidence of a wave is equal to the angle of reflection of the wave as measured from a line normal (perpendicular) to the barrier. This simply means that the wave will bounce off at the same angle at which it came in.

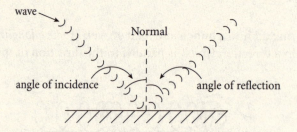

Refraction is the bending of a wave due to a change in medium. A water wave moving from deep water to shallow water will bend its path, slow down, and shorten its wavelength.

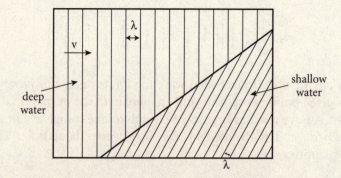

The speed of the wave and its wavelength always change when a wave undergoes refraction, but its frequency does not change. According to the equation $v = f\lambda$, a slower speed means having a shorter wavelength, and a higher speed means having a longer wavelength.

Diffraction is the bending of a wave around a barrier, such as water waves bending around a rock in a stream, or sound waves bending around the corner of a building. When water waves pass through a narrow opening, the sides of the waves drag on the walls of the opening causing these parts of the wave to lag behind the center of the wave and creating a semi-circular wave pattern, as shown in the figure below.

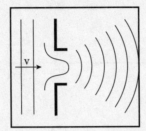

When two waves are traveling in the same medium at the same time, they *interfere* with each other. For example, consider two waves moving toward each other in the same rope. If the waves are moving on the same side of the rope, they build on each other and create a larger wave when they occupy the same space at the same time. This is called *constructive interference*, and the large wave produced at that instant is called an *antinode*.

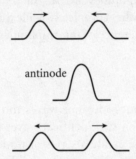

The addition of the displacements of two or more waves at each point in a medium is called the *principle of superposition*.

We say that the waves are *in phase* when they interfere constructively. After the waves pass through each other, they continue moving as if they never interfered.

If two waves of equal amplitude approach each other on opposite sides of the rope, they interfere destructively, and the waves destroy each other for the instant they are occupying the same space, and a node is created at that point. A node is a point of no displacement, in this case resulting in a flat rope.

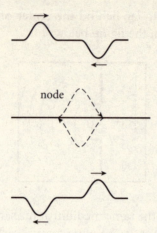

We say that the waves are *out of phase* when they interfere destructively. Once again, after the waves pass through each other, they continue moving as if they had never interfered.

Of course, two waves of different amplitudes also can interfere destructively. If they have different amplitudes, they won't cancel each other completely while they are interfering, but the amplitude of the resultant wave will be the difference of the amplitudes of the two interfering waves.

Standing Waves

The term *standing wave* is an oxymoron, since waves must move and never stand still. But waves can appear to stand still when two identical waves traveling in opposite directions in the same medium at the same time create a series of nodes and antinodes. Consider a rope tied to a wall. If we send a wave down the rope toward the wall, the wave reflects off the wall on the opposite side from which it was sent, according to the law of reflection.

If we continue to send regular waves down the rope and they continue to reflect off the wall, the incident and reflected waves will reinforce each other in some places and cancel each other in other places. The result is a series of antinodes (loops) where constructive interference is occurring, and nodes (points of no displacement between the loops) where destructive interference is occurring. We call the pattern produced a standing wave.

The pattern above, called a harmonic or overtone, shows three antinodes (loops) and occupies $\frac{3}{2}$ of a wavelength.

Another type of standing wave is produced when water waves are passed through two openings, called a double-slit, and an interference pattern results. The semi-circular wave patterns which emerge from the slits interfere with each other, creating nodes and antinodes, and we see a pattern like the one in the figure below:

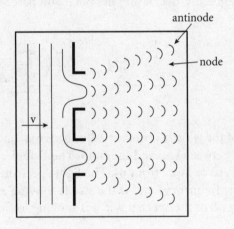

Sound Waves and the Doppler Effect

Sound is a mechanical longitudinal wave, and therefore must have a medium to travel through. Sound generally travels at about 340 m/s in air, but it travels at considerably higher speeds in more dense media such as water or steel. The characteristics of sound which are produced and how we detect and perceive these characteristics are summarized in the table below.

Sound produced as:	Sound detected as:
Frequency	Pitch
Amplitude	Loudness or volume
Harmonics	Quality or tone

The third characteristic, harmonics, is the combination of several simultaneous frequencies that give a sound its special tone. For example, we can tell the difference between a trumpet and a clarinet because our ear detects the special harmonics, even if they are playing the same pitch at the same loudness. For the same reason, we can tell the difference between two voices.

When a sound source is moving toward you, you hear a slightly higher pitch than if the sound source is at rest relative to you. By the same token, when a sound source is moving away from you, you hear a slightly lower pitch. This phenomenon is called the *Doppler effect*. For example, if a train is traveling toward you while blowing its horn at a certain pitch, the waves will appear to be arriving at your ear more frequently, increasing the pitch you perceive.

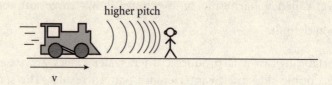

If the train blows its whistle while accelerating away from you, the waves will appear to be arriving at your ear less frequently, decreasing the pitch you perceive.

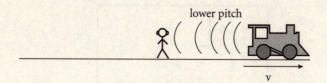

Of course, the frequency of the whistle is not actually changing, but you perceive it to change due to the relative motion between you and the train. The Doppler effect also occurs when a light source is moving toward or away from us. The light spectrum of a star, for example, is shifted toward the red (low frequency) end if the star is moving away from us, and toward the blue (high frequency) end of the spectrum if it is moving toward us.

Standing Sound Waves in a Closed Pipe

A closed pipe is one in which one end of the pipe is closed and the other end is open. If we send a sound wave into the pipe and let it reflect off of the closed end, it will return to the top of the pipe, as shown in the figure below:

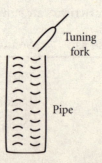

If we continually send sound waves down into the pipe to match the rate at which they are being reflected off the closed end, we will set up a *resonance condition*, that is, a standing wave. If the length of the pipe is such that a whole number of quarter-wavelengths fits into it, we will hear loud tones called overtones. In other words, there must be a node at the closed end of the pipe and an antinode at the open end. The lengths of the pipes which fit the first three overtones are sketched below:

Physics Review
WAVES AND OPTICS

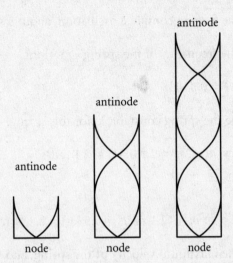

If we know the frequency of the sound waves and the length of the pipe, we can find their wavelength and then their speed by $v = f\lambda$.

Since light is a wave, we will apply all of the characteristics of waves that we have discussed to the characteristics of light in the next few sections.

Sample Problem:

The following is a plot of the motion of a 0.5 kg mass oscillating on a spring.

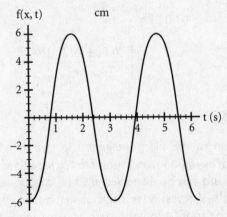

(a) What is the amplitude of the spring's motion?

Answer: Reading from the graph, the oscillator's amplitude is maximum displacement from equilibrium, which is 6 cm.

(b) What is the period of the spring's motion?

AP Physics

CHAPTER 8

Answer: The period is time for one complete oscillation, about 3 seconds.

(c)   What is the frequency of the spring's motion?

Answer: Frequency, $f = 1/T = 0.33$ Hz

(d)   Determine the spring constant, k, for this spring.

Answer: Angular frequency, $\omega = 2\pi f = 2\pi(0.33) = 2.1$ rad/s

Since $\omega = (k/m)^{1/2}$

$$k = \omega^2 m = (2.1 \text{ rad/s})^2(0.5 \text{ kg}) = 2.2 \text{ N/m}$$

(e)   What is the maximum velocity of the spring, and when during the oscillation does it occur?

Answer: There are a couple of ways to determine this. One way would be to write the position versus time equation for the oscillation and take the first derivative to find the velocity amplitude directly. Using the energy method (and no calculus), first find the maximum energy of the oscillator:

$$U = \frac{1}{2} kA^2 = (\frac{1}{2})(2.2 \text{ N/m})(0.06 \text{ m})^2 = 0.004 \text{ J}$$

Then convert this maximum potential energy at the spring's amplitude to maximum kinetic energy, which occurs at equilibrium:

$$0.004 \text{ J} = \frac{1}{2}mv^2$$

$$v = \left[\frac{(0.004 \text{ J})(2)}{(0.5 \text{ kg})} \right]^{1/2} = 0.13 \text{ m/s}$$

II. PHYSICAL OPTICS

Light is the visible portion of the *electromagnetic spectrum*, which includes many other electromagnetic waves such as *radio waves, ultraviolet light,* and *x-rays*. Visible light travels at 3×10^8 m/s in a vacuum, and can be dispersed into its component colors, with the longest wavelength being red and the shortest wavelength violet. Since light is a wave, all of the terms we apply to any waves can be applied to light, such as *wavelength, refraction,* and *interference*.

Key Terms

antinode: point of maximum displacement of two or more waves constructively interfering

constructive interference: addition of two or more waves which are in phase, resulting in a wave of increased amplitude

WAVES AND OPTICS

destructive interference: addition of two or more waves which are out of phase resulting in a wave of decreased amplitude

diffraction: the spreading of a wave beyond the edge of a barrier or through an opening

diffraction grating: material containing many parallel lines which are very closely spaced so that when light is passed through the lines, an interference pattern is produced

dispersion of light: the spreading of light into its component colors, usually by a prism or diffraction grating

electromagnetic wave: a wave which is produced by vibrating charges and propagates itself through space by the mutual generation of changing electric and magnetic fields

gamma ray: high energy electromagnetic wave emitted by a radioactive nucleus

infrared: electromagnetic waves of frequencies just below those of red visible light

laser: light amplification by stimulated emission of radiation; laser light is coherent, that is, all the wavelengths are in phase with each other

monochromatic light: light having a single color or frequency

node: the point of no displacement in a standing wave

primary light colors: red, green, or blue light

opaque: material that will not transmit light

polarized light: light in which the electric fields are all in the same plane

translucent: a material which passes light but distorts its path

transparent: material through which light can pass without distorting the direction of the rays

ultraviolet: electromagnetic waves of frequencies higher than those of violet light

spectrum: the range of electromagnetic waves from low frequency to high frequency, or colors when white light is passed through a prism

speed of light: in a vacuum, 3×10^8 m/s

white light: visible light consisting of all colors

x-ray: high-frequency, high-energy electromagnetics waves or photons

Electromagnetic Waves

As we briefly discussed in the last chapter, an electromagnetic wave is a vibration of electric and magnetic fields that move through space at an extremely high speed. The electric and magnetic fields in an electromagnetic wave vibrate perpendicular to each other. The electromagnetic wave spectrum, listed from lowest frequency to highest frequency, include *radio waves, microwaves, infrared, visible light, ultraviolet, x-rays, and gamma rays*. If the *visible colors of light are listed from long wavelength (low frequency) to short wavelength (high frequency), they would follow the order red, orange, yellow, green, blue, and violet (ROYGBV)*.

All electromagnetic waves are a result of the same phenomena, and although they have different wavelengths and frequencies, they all travel through a vacuum at exactly the same speed: 3×10^8 m/s, or about 670 *million miles per hour*. This speed is often referred to as the speed of light, although light is just one example of an electromagnetic wave. More accurately, this speed is the speed of any electromagnetic wave in a vacuum. In any case, the speed of an electromagnetic wave is given the symbol c, from the Latin word *celeritas*, meaning "swift."

Polarization

Light is a transverse vibration of electric and magnetic fields. If you could watch light waves coming toward you, you would see that they actually vibrate in many directions. We say the light is *polarized* when it is forced to vibrate in only one plane:

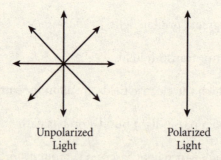

Unpolarized Light   Polarized Light

Diffraction and Interference

As discussed in an earlier section, diffraction is the bending of a wave around a barrier or through an opening. If we pass a light wave through a narrow single slit, it will behave very similarly to the water waves discussed earlier, with the edges of the light waves lagging behind the center of the waves. If we were to place a screen opposite from the single-slit opening, we would see a bright light near the center of the screen with narrow lines becoming dimmer toward the edges of the screen. These lines form because of interference from light waves arriving from different locations within the slit.

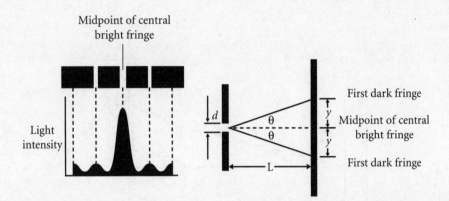

If we replace the single-slit opening with a double-slit opening, the pattern on the screen changes. As the light passes through the two openings, it becomes two sources of light waves instead of one. These two light waves behave like the semi-circular water waves we observed earlier, interfering constructively in some places and destructively in others. The pattern on the screen would consist of a central bright band of light, with alternating light and dark bands toward the edges of the screen:

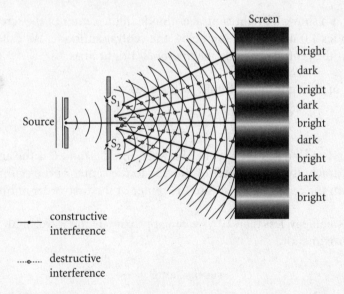

The bright bands on the screen are the places where constructive interference is occurring (*antinodes*) and the dark bands are a result of destructive interference (*nodes*). In 1801, Thomas Young was able to measure the wavelength of light waves using this double-slit diffraction pattern. He found that for a given distance between the slits and length from the slits to the screen, the width of the bright central antinode on the screen is proportional to the wavelength of the light. Thus, red light would produce a wider bright central band than violet light.

Consider the monochromatic (one color) light passing through a double-slit and creating an interference pattern of nodes and antinodes on the screen at the right. The pattern of light on the screen is represented by the graph of light intensity I as a function of position y.

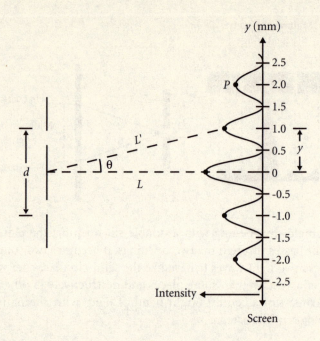

The intensity graph shows a bright central antinode in the center of the screen, and the next brightest antinodes 1.0 mm above and below the central antinode. We call these antinodes above and below the central antinode the *first-order* bright lines.

The wavelength of the light can be found by the equation

$$m\lambda = d \sin \theta$$

where $m = 1$ (first order), d is the distance between the slits, and θ is the angle between the line of length L drawn from the center of the slits to the center of the central antinode, and the length L' from the center of the slits to the center of the first-order antinode.

If the angle θ is small, say, less than 10°, we can approximate the wavelength by assuming the small-angle approximation:

$$\sin \theta \approx \tan \theta = \frac{y}{L}$$

Then:

$$\lambda = d \sin \theta = d\left(\frac{y}{L}\right)$$

We might call this equation a *first-order approximation* of the wavelength of the light.

If we consider the point P on the screen, we see that it is located at the second-order bright line, or $m = 2$. A line drawn from the lower slit to point P differs in path length from a line drawn from the upper slit to point P by a *path difference* of $n\lambda$, or in this case 2λ. If m is a whole number, the light rays from the two slits will interfere constructively, and if m is not a whole number, such as $m = \frac{5}{2}$ at point Q on the screen, the light rays will interfere destructively.

(Note: This author prefers to use m, rather than n, for order of diffraction to avoid confusion with n as index of refraction in some optics problems.)

SAMPLE PROBLEM:

Blue monochromatic light is shone through a double slit apparatus in which the slit separation, d, is 0.001 m.

(a) What is the distance between the third order bright fringes on either side of the centerline of the screen if the screen is 1.5 meters from the slits?

Answer:

The third order bright fringe is noted by setting $m = 3$ and using the equation:

$$m\lambda = d \sin \theta$$

Assume a value of 400 nm for the wavelength of blue light.

Thus, the angle from the central line to the third bright fringe is:

$$\theta = \sin^{-1}\left(\frac{m\lambda}{d}\right) = \sin^{-1}\left[3\left(\frac{400 \times 10^{-9}}{.001}\right)\right] = .069 \text{ degrees}$$

To find the distance from the center to this line, we use simple trigonometry:

$$\tan .069° = \frac{x}{1.5}, \text{ so } x = 1.5 \tan(.069°) = .0018 \text{ m, or } 1.8 \text{ mm.}$$

Thus, the total distance between third order fringes is 2×1.8 mm = 3.6 mm.

Alternate Solution: Use the small angle approximation, which is

$$x \approx \frac{m\lambda L}{d}$$

$$x = \frac{(3)(400 \times 10^{-9})(1.5 \text{ m})}{0.001} = 0.0018$$

Thus, the distance from the third order bright line on one side to the other side is 0.0018 doubled, or 0.0036 m.

(b) If the double slit were replaced with a diffraction grating with 1,400 lines/inch, would the distance described in part *a* increase or decrease, and by how much?

Answer:

When we replace the double slit with a diffraction grating, the only part that really changes is the value of d. Now, instead of having two slits to measure, we have several hundred. We need to know the separation of any two slits, so we take the value for "lines per inch" and turn it around to make it "meters per line."

1 inch = .0254 m, so 1,400 lines per inch is:

$$\frac{.0254 \text{ m}}{1,400 \text{ lines}}, \text{ and } d = 1.8 \cdot 10^{-5} \text{ m}.$$

Substituting into the same equation, we find that $\sin \theta_m = \frac{(m\lambda)}{d}$.

Now, $\sin \theta_m = \frac{3(400 \cdot 10^{-9})}{1.8 \cdot 10^{-5}} = .067$, and $\theta_m = $ approx 0.067 radians or 3.82 degrees.

Converting the angle to distance from center, we find that:

$$\tan .067^R = \tan 3.82° = \frac{x}{1.5}, \text{ and } x = 0.10 \text{ m}$$

Doubling that to find the total separation distance, we get a distance of 0.20 m. That **is** a significant change from the double-slit problem.

Our solution, then, is to say that the distance increases by 196.4 mm.

III. GEOMETRIC OPTICS

The ray model of light states that light may be represented by a straight line along the direction of motion, and *ray optics* is the study of light using the ray model. Light can be *reflected* from a surface such as a *mirror*, and *refracted* through a transparent medium such as a *lens*. The relationship between the speed of light in two different media and the angle of the light rays can be found using *Snell's law of refraction*. As light rays are reflected or refracted, they may form an *image*. Images can be *real* or *virtual*, depending on the distance from the lens or mirror to the *object* which is the source of the light rays. *Ray diagrams* can be drawn to show the bending of a light ray or to locate an image formed by a lens or mirror.

Key Terms

angle of incidence: the angle between the normal line to a surface and the incident ray or wave

angle of reflection: the angle between the normal line to a surface and the reflected ray or wave

angle of refraction: the angle between the normal line to a surface and the refracted ray or wave at the boundary between two media.

converging lens: a lens which converges light rays to a focal point; also known as a convex lens

converging mirror: a mirror which converges light rays reflecting from it; also known as a concave mirror

WAVES AND OPTICS

critical angle: the minimum angle entering a different medium at which total internal reflection will occur

diffuse reflection: reflection of light in many directions by a rough surface

diverging lens: a lens which diverges light rays passing through it; also known as a concave lens

diverging mirror: a mirror which diverges light rays reflecting from it; also known as a convex mirror

focal length: the distance between the center of a lens or mirror to the point at which the rays converge at the focal point

focal point: the point at which light rays converge or appear to originate

image: reproduction of an object using lenses or mirrors

index of refraction: the ratio of the speed of light in a vacuum to the speed of light in another medium

lens: a piece of transparent material that can bend light rays to converge or diverge

magnification: ratio of the size of an optical image to the size of the object

object (**optics**): the source of diverging light rays

plane mirror: smooth, flat surface that reflects light regularly

principal axis: the line connecting the center of curvature of a curved mirror with its geometrical vertex; the line perpendicular to the plane of a lens passing through its center

ray model of light: light may be represented by a straight line along the direction of motion

ray optics: the study of light using the ray model

real image: an image that can be projected onto a screen

refraction: the change in speed, wavelength, and direction of a light ray due to a change in medium

total internal reflection: the complete reflection of light that strikes the boundary between two media at an angle greater than the critical angle

virtual image: an image which cannot be projected onto a screen; point at which diverging light rays appear to originate

Reflection of Light

Any wave that bounces off of a barrier follows the *law of reflection*: the angle of incidence is equal to the angle of reflection as measured from a line normal (perpendicular) to the barrier. In the case of light, the barrier is often a mirror.

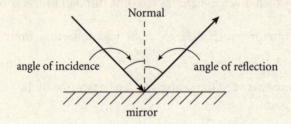

This is why if someone can see you in a mirror, you can see him in the mirror as well.

Refraction of Light

If you put a pencil in a clear glass of water, the image of the pencil in the water appears to be bent and distorted. The light passing from the air into the water is *refracted*, bending due to the fact that it's passing from one medium to another. If we consider a single beam of laser light, we can observe it as it passes from air into a piece of glass.

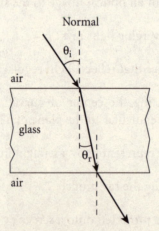

The angle θ_i from the normal line at which the beam approaches the glass from the air is called the *angle of incidence*. The angle θ_r from the normal line in the glass is the *angle of refraction*. As the light passes from the air, a less dense medium, into the glass, a more dense medium, the beam bends toward the normal line. When the beam of light exits the glass and passes back into the air, it bends away from the normal at the same angle it entered the glass from the air.

The light bends toward the normal in the glass because the beam slows down as it enters the glass. Light travels more slowly in a more dense medium. Recall that sound travels faster in a more dense medium, but sound is a mechanical wave, while light is an electromagnetic wave.

The ratio of the speed of light in air (approximately a vacuum) to the speed of light in the glass (or any other medium) is called the *index of refraction n*:

$$n = \frac{c}{v_{glass}}$$

For example, the index of refraction for a diamond is about 2.5, which means that light travels 2.5 times faster in a vacuum than in diamond.

Prism Dispersion

It turns out that each color in the spectrum refracts just a little differently than every other color. This is why we can separate white light into its component colors by passing it through a prism. The shorter wavelengths slow down and bend more than the longer wavelengths, so violet bends the most, and red the least:

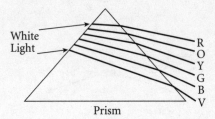

Snell's Law of Refraction

We can relate the index of refraction to the angles of incidence and refraction by using *Snell's law of refraction*:

$$n_1 \sin \theta_1 = n_2 \sin \theta_2$$

where n_1 and n_2 are the indices of refraction of the first and second media, and θ_1 and θ_2 are the angles of incidence and refraction, respectively.

Total Internal Reflection

Consider a waterproof laser that you can put underwater and shine a beam of light up out of the water into the air. If you shine the light at a small angle relative to the normal, the light will emerge from the water and bend away from the normal as it enters the air.

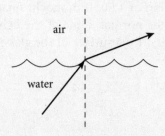

As you increase the angle at which the laser is pointed at the surface of the water, the refracted angle also increases, eventually causing the refracted ray to bend parallel to the surface of the water:

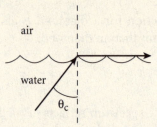

The angle θ_c is called the *critical angle*. If the laser is pointed at an angle greater than the critical angle, the beam will not emerge from the water, but will reflect back into the water:

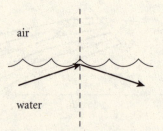

This phenomenon is called *total internal reflection*. The inside surfaces of a glass prism in a pair of binoculars can become like mirrors, reflecting light inside the prism if the light is pointed at the surface at an angle greater than the critical angle. Total internal reflection is the also the principle behind the transmitting of light waves through transparent fiber optic cable for communication purposes.

Thin Film Interference

If you hold a compact disc (CD) at an angle to a light source, you see colors of light reflecting from the surface of the CD. Different angles of incidence reflect different colors. There is a thin transparent film on the reflecting (mirrored) surface of the CD which separates the colors. But the colors that are reflected depend on the refraction, reflection, and interference of the light.

Consider a thin film with an index of refraction of 1.40 and thickness, t, that is on the surface of glass with an index of refraction of 1.50. If monochromatic light is incident on the film from the air at a small angle to the normal, some of the light is reflected, and some of it is refracted in the film and reflected off the surface of the glass:

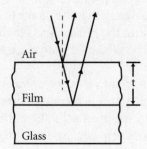

The ray that is refracted in the film eventually emerges parallel to the ray reflected off the surface of the film, but has traveled a longer distance. If the incident light enters and exits the film perpendicular to the surface of the film, then the extra distance traveled by the light ray in the film is twice the thickness of the film, or 2t. The two emerging rays can interfere with each other. If they are *in phase*, that is, they differ by a whole number of wavelengths in the film ($1\lambda, 2\lambda, 3\lambda, \ldots$), then they will interfere constructively, and your eye would see a bright light (antinode) reflected back into the air. If the two emerging rays are *out of phase*, that is, they differ by a half number of wavelengths in the film ($\frac{1}{2}\lambda, \frac{3}{2}\lambda\ldots$), they will interfere destructively, and the light reflected back to the air would be a minimum (node). Whether the light rays are a whole or half number of wavelengths, in or out of phase, depends on how much farther the refracted ray has to travel through the film, which would be 2t. Thus, the condition for interference is

$$2t = m\lambda$$

where $m = 1, 2, 3, \ldots$ for *constructive* interference and $m = \frac{1}{2}\lambda, \frac{3}{2}\lambda$, for *destructive* interference. Remember, the wavelength in the equation above is the wavelength in the film, which can be found by

$$\lambda_{film} = \frac{\lambda_{air}}{n_{film}}$$

where n_{film} is the index of refraction of the film.

Mirrors

We will discuss three types of mirrors in this chapter: *plane*, *diverging*, and *converging*.

A *plane mirror* is simply a flat mirror. From your everyday experience you know that a plane mirror always produces an image which is the same size as the object (which could be you in the morning), left-right reversed, and the same distance behind the surface of the mirror as the object is in front of the mirror. We say that the image formed by a plane mirror is *virtual*, since we cannot place a screen behind the mirror and see the image projected on the screen. A virtual image is one that cannot be projected onto a screen.

A *diverging mirror* is sometimes referred to as a *convex mirror*. You may have seen this type of mirror in the corner of a convenience store. The mirror diverges the rays of light which strike it, allowing the clerk at the store to see practically the entire store in one mirror. Light rays coming into the mirror parallel to its principal axis will diverge, or spread apart:

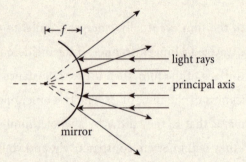

Notice that the rays appear to originate from a point behind the mirror. This point is called the *virtual focus*, and the distance between the surface of the mirror at its center and the focal point is called the *focal length f*.

A *converging mirror* is sometimes referred to as a *concave mirror*. If a spherically-shaped concave mirror is small compared to its radius of curvature *r*, then light rays coming in parallel to the principal axis of the mirror will converge to a focal point.

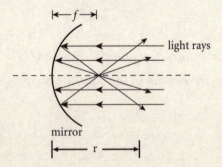

For this kind of mirror, the focal length f and the radius of curvature r of the mirror are related by the equation $f = \frac{1}{2} r$.

Images Formed by a Converging (Concave) Mirror

If you look into a converging mirror, you will at first see an image of yourself which is inverted (upside-down), but then as you move closer to the mirror you will see your image turn upright as you pass the focal point of the mirror. Satellite dishes act as converging mirrors for radio and TV waves, gathering them at a detector located at the focal point of the dish. Most research telescopes also use converging mirrors rather than lenses to initially focus incoming light and study images.

The image formed by a converging mirror depends on the location of the object in relation to the focal length and its orientation, and can be *real* (which means it can be projected onto a screen), or virtual. The image can also be *upright* or *inverted* (upside-down), and *smaller*, *larger*, or the *same size* as the object.

For example, if we choose a burning candle as our object, we can place it on the principal axis of the converging mirror. Let's place the candle at an object distance d_o that is greater than twice the focal length of the mirror:

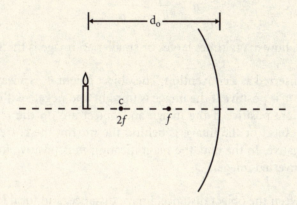

To find the image formed by the mirror, we will draw two rays: one ray from the flame that strikes the mirror parallel to the principal axis and reflects through the focal point, and another ray from the flame that goes through the focal point and reflects back parallel to the principal axis. The image is formed at the location of the intersection of these two rays:

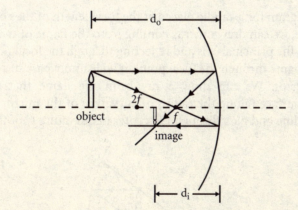

The object distance d_o is the distance from the center of the surface of the mirror to the object, and the image distance d_i is the distance from the center of the surface of the mirror to the image. Note that when the object is place at a distance greater than twice the focal length, the image is inverted and smaller than the object. If we place a screen at the location of the image, we would see that the image is real. For a concave mirror, if the image is formed on the same side of the mirror as the object, then the image is real. If the image is formed on the opposite side of the mirror as the object, it is virtual. Recall that a plane mirror will only produce an image on the opposite side to the object, and thus is always virtual.

For the image formed, the ratio of the object height h_o to the image height h_i is equal to the ratio of the object distance to the image distance:

$$\frac{h_o}{h_i} = \frac{d_o}{d_i}$$

The magnification of the image is given by:

$$m = \frac{h_1}{h_o} = -\frac{d_i}{d_o}.$$

Magnification tells us how many times larger or smaller an image is than the object.

The negative sign is inserted as a convention. The object height h_o is always taken as positive, and the image height h_i is positive if the image is upright and negative if inverted. The object and image distances are positive if the image and object are on the reflecting side of the mirror. If either the object or the image is behind the mirror, the corresponding object or image distance is negative. In the end, the magnification m is positive for an upright image and negative for an inverted image.

The relationship between the object distance, image distance, and focal length is

$$\frac{1}{f} = \frac{1}{d_o} + \frac{1}{d_i}$$

where $f = \frac{1}{2}r$.

Let's draw the ray diagram for a candle placed at the focal length of the concave mirror. As in the previous example, we can draw one ray coming from the flame of the candle that strikes the mirror parallel to the principal axis and reflecting through the focal point. However, if we draw a ray from the flame through the focal point, it will simply pass straight downward and never strike the mirror. We can draw a ray from the flame that strikes the mirror perpendicular to its surface (along the radius of curvature of the mirror), and it will reflect back along the same line and pass through the center of curvature C of the lens.

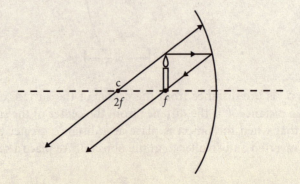

We see that the rays coming from a candle placed at the focal point of a concave mirror do not intersect each other and thus no image of the candle is formed.

If we place the candle at a distance less than the focal length, the reflected rays diverge, and the image is formed at the point from which the rays seem to originate:

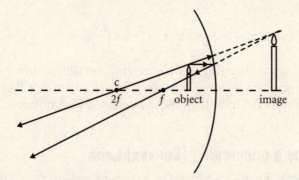

The image formed by a concave mirror when the object is placed inside the focal point is virtual, upright, and enlarged. Thus, if you want to see an upright image of your face in a concave mirror, you must move the mirror to a distance less than one focal length to your face.

A summary of the images formed by a concave mirror (and a convex lens) is listed in the table near the end of this section.

Lenses

Lenses operate on the principle of refraction. A *diverging lens* is a lens that is thicker on the edges than it is in the middle, and it diverges the light rays that pass through it:

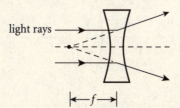

A diverging lens is sometimes referred to as a *concave lens*. The focal point of a diverging lens can be found by extending the diverging rays back behind the lens until they seem to meet.

A *converging lens* is a lens that is thicker in the middle than on the edges, and it converges parallel rays that pass through it:

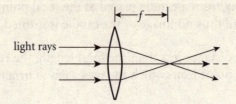

A converging lens is sometimes referred to as a *convex lens*.

When you read the words *diverging* and *converging*, or *convex* and *concave*, be sure you identify whether the question is asking you about lenses or mirrors. The answers associated with lenses might be quite different than those associated with mirrors!

Images Formed by a Converging (Convex) Lens

Let's take a closer look at the images formed by a converging lens. We said earlier that a *virtual image* is one that cannot be projected onto a screen. On the other hand, a *real image* is one that *can* be projected onto a screen. If your teacher uses an overhead projector in your classroom, the image projected on the screen is a real image. The image projected on a screen at a movie theater is also real. Under certain circumstances, a converging lens can create an image that can be real or virtual, upright or inverted, enlarged or reduced in size, or the same size as the object. It all depends on where the object is placed relative to the focal length of the lens.

Let's choose a burning candle for our object, since it produces light from the flame at the top of the candle. We'll draw a side view of a converging lens, include a horizontal principal axis through the center of the lens, and mark the focal length of the lens on either side of it. Let's place a candle at a distance greater than twice the focal length from the center of the lens:

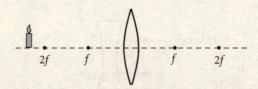

To find out what kind of image will be formed by the lens, we will draw two rays: one ray from the flame entering the lens parallel to the principal axis and bending through the focal point, and another ray from the flame which passes straight through the center of the lens without bending. The image is formed at the location of the intersection of these two rays:

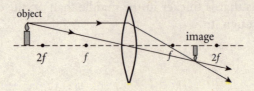

We see that in the case where the object (candle) distance from the lens is greater than twice the focal length, the image is inverted and reduced in size. The image is also real, so if we placed a screen at the location of the image, we would see the projection of a small inverted candle. The image formed by a converging lens is real if the object distance is greater than the focal length.

If we placed the candle at a distance from the lens equal to the focal length, our two rays would emerge parallel to each other, and no image would be formed:

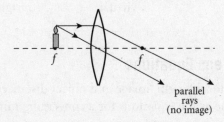

If we placed the candle inside the focal length, our two rays would diverge as they emerged from the lens. No image would be formed on the side opposite to the candle, but if we extended the rays backward, we would find that they seem to originate on the same side as the candle. The point from which they seem to originate is where a virtual image of the candle is formed:

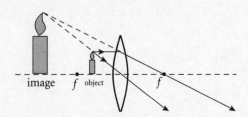

A summary of the images formed by a converging lens and a converging mirror is listed in the table below, where d_o is the distance from the candle to the object and f is the focal length of the lens or mirror. For a converging lens, a positive (+) image distance d_i implies that the image is formed on the opposite side of the lens as the object, and a negative (−) image distance implies that the image is formed on the same side as the object. This sign convention is just the opposite for a converging (concave) mirror.

Object placed: at:	Image distance d_i	real or virtual	upright or inverted	enlarged or reduced
$d_o > 2f$	+	real	inverted	reduced
$d_o = 2f$	+	real	inverted	same size
$f < d_o < 2f$	+	real	inverted	enlarged
$d_o = f$	No image	No image	No image	No image
$d_o < f$	–	virtual	upright	enlarged

Converging (Convex) Lens Equations

The equations relating the focal length, image and object distances, and magnification for a converging lens are the same as the equations for a converging mirror:

$$\frac{1}{f} = \frac{1}{d_o} + \frac{1}{d_i}$$

The magnification of the image is given by:

$$m = \frac{h_i}{h_o} = -\frac{d_i}{d_o}$$

Sample Problem:

Find the final location, size, and type of image formed when the object is placed, as shown in the drawing below, 30 cm in front of a convex lens, which is 1.0 m from a convex mirror.

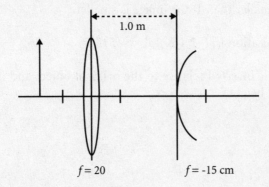

object distance = 30 cm
object height = 10 cm

Answer:

First, use the thin lens equation to determine the properties of the image formed by the lens:

$$\frac{1}{f} = \frac{1}{d_I} + \frac{1}{d_o}$$

$$\frac{1}{20} = \frac{1}{d_I} + \frac{1}{30}$$

$$d_I = 60 \text{ cm}$$

The positive image distance indicates that the initial converging lens will produce a real, inverted image to the right of the lens. This places the image 40 cm in front of the convex mirror.

Using 40 cm as the object distance for the mirror and –15 cm as the focal length, we find that the final image distance equals:

$$\frac{1}{(-15)} = \frac{1}{40} + \frac{1}{d_i}$$

$$d_I = -10.9 \text{ cm}$$

This places the final image behind the mirror, between the mirror and its focus. The image is virtual and oriented the same as the first image, which is inverted.

To find the magnification we need to find both magnifications and multiply them together.

$$m_1 = -\frac{60}{30} = -2$$

This is for the real, inverted image formed by the lens.

$$m_2 = -\frac{(-10.9)}{40} = .2725$$

The positive magnification means that the image is oriented the same as the object. Because the magnification is smaller than 1, the image is smaller.

The combined magnification is: $(-2)(.2725) = -.545$

Thus, the final image is inverted relative to the original object and is just over half as large (54.5% of the original height).

Physics Review

WAVES AND OPTICS

Important Equations

Hooke's law for a spring:	$\mathbf{F}_s = -k\mathbf{x}$
Potential energy of a spring:	$U = \left(\dfrac{1}{2}\right)kx^2$
Displacement as a function of time for an object in SHM:	$x(t) = A\sin(\omega\tau)$
Period of an object in SHM:	$T = 2\pi\left(\dfrac{m}{k}\right)^{1/2}$
	$T = \dfrac{2\pi}{\omega}$
Period of a pendulum:	$T = 2\pi\left(\dfrac{L}{g}\right)^{1\backslash 2}$
Velocity as a function of time for an object in SHM:	$v(t) = A\omega\cos(\omega t)$
Acceleration as a function of time for an object in SHM:	$a(t) = -A\omega^2\sin(\omega t)$
Speed of a wave:	$v = f\lambda$
Speed of a wave on a string:	$v = \left[\dfrac{T}{(m/L)}\right]^{1/2}$
Natural frequencies of a string:	$f_n = \dfrac{(nv)}{(2L)}$ for $n = 1, 2,$
Speed of sound in air as a function of temperature:	$v_{air}\ (m/s) = 331 + 0.6\ T\ (°C)$
Destructive interference condition for two sound waves:	$\Delta L = (n\lambda + \dfrac{\lambda}{2}),\ n = 0, \pm 1, \pm 2$
Resonance lengths for an open pipe:	$L = \dfrac{n\lambda}{2};\ n = 1, 2, 3, ...$
Resonance lengths for a closed pipe:	$L = \dfrac{m\lambda}{4};\ m = 1, 3, 5, ...$

KAPLAN 225

AP Physics
CHAPTER 8

Relationship between frequency and wavelength for electromagnetic waves:	$c = \lambda f$
Index of refraction:	$n = \dfrac{c}{v}$
Law of reflection:	$\theta_i = \theta_r$
Relationship between the radius of curvature and the focal length:	$R = 2f$
Mirror equation:	$\dfrac{1}{d_i} + \dfrac{1}{d_o} = \dfrac{1}{f}$
Magnification equation:	$M = -\dfrac{d_i}{d_o} = \dfrac{h_i}{h_o}$
Snell's law:	$n_1 \sin \theta_1 = n_2 \sin \theta_2$
Lens equation:	$\dfrac{1}{f} = \dfrac{1}{d_i} + \dfrac{1}{d_o}$
Critical angle:	$\sin \theta_c = \dfrac{n_2}{n_1}$
Diffraction equations:	$m\lambda = d \sin \theta$
Small angle diffraction:	$x_m \approx \dfrac{m\lambda L}{d}$

PRACTICE PROBLEMS

1. Use this graph of displacement versus time to answer parts *a* through *d* below:

   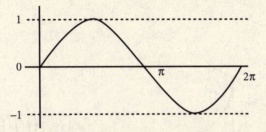

   (a) Sketch the graph of velocity versus time for this function on the axes below:

   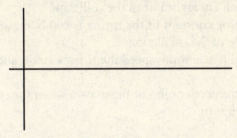

   (b) At what time (*t*) is the velocity equal to zero?
   (c) Sketch the graph of acceleration versus time for this function
   (d) At what time (*t*) is the acceleration equal to zero?

2. A mass slides down a frictionless ramp as shown and collides with another stationary mass that is attached to a spring as shown below. The two masses stick together and enter into Simple Harmonic Motion.

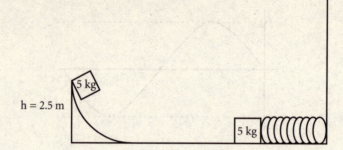

   (a) What is the total momentum of the two masses just after they collide?
   (b) How much energy is lost in the collision?
   (c) If the spring constant of the spring is 200 N/m, what will be the amplitude of the oscillation?
   (d) What is the maximum speed the masses attain and where will this occur?
   (e) What is the acceleration of the masses when the spring compression is a maximum?

3. Yellow light of wavelength 580 nm shines on a diffraction grating with 4,000 lines/cm.

   (a) At what angles do the first-order and second-order maxima occur?
   (b) In order to study the structure of crystals, scientists normally use X-rays. Why do they use X-rays and not the visible light of part *a*?

4. For each part, label appropriate distances, indices of refraction, and angles.

   (a) Draw a ray diagram that shows how a convex mirror produces a virtual image.
   (b) Draw a ray diagram that shows how a concave mirror produces a real image that is larger than the object.
   (c) Show the path of a ray of light as it travels from a more dense medium to a less dense medium.

5. Use the diagram of the compound microscope shown below to answer the following:

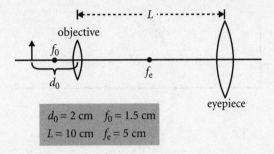

(a) Use a ray diagram to locate the image.
(b) On your diagram, indicate the image formed by the first lens, and the object for the second lens.
(c) Find the magnification of the microscope.
(d) Describe the major difference between this microscope and a telescope, which also uses two lenses in combination.

6. The distance between the movable mirror and the beam splitter in a Michelson interferometer is increased a small amount. When this happens, you see 200 dark fringes move across the field of view. If the incident light was 600 nm, by how much was the mirror moved (in millimeters)?

7. Blue light (λ = 450 nm) shines on a diffraction grating that has 4,000 lines/inch.

(a) Calculate the angle of the fourth-order image.
(b) For this grating and light source, can you have a 15th order image?

8. Consider the arrow shown here in front of a concave mirror, approximately halfway between the focal point and the center of curvature (both marked on the mirror axis by small black dots). Construct the image for this situation and state the value for the magnification.

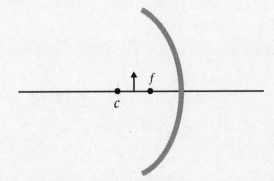

9. Construct an image for the arrow shown below, which is outside the center of curvature of the convex mirror, and state if this image is virtual or real.

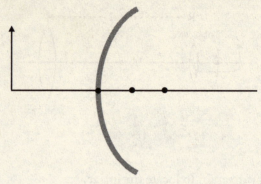

10. (a) On the diagram below, sketch the interference pattern that would be produced by a beam of green light passing through a double slit of width 0.002 m.

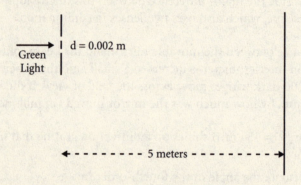

(b) Indicate the locations of the central bright fringe and the first two maxima on either side of the center, including their distance from the central bright fringe.

(c) Describe and sketch the pattern that would be produced if the double slit were replaced with a diffraction grating.

Physics Review

WAVES AND OPTICS

Turn page for Solutions to Practice Problems ➡

KAPLAN 231

Solutions to Practice Problems

1. Answers:

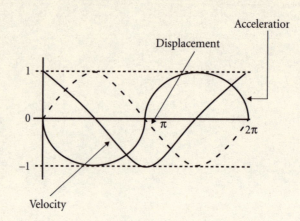

(a), (c) The position function is a sine curve with amplitude 1 and period 2π. The frequency is $\frac{1}{2}\pi$, so using the form $x(t) = A \sin 2\pi ft$, the position function has the equation:

$$x(t) = (1) \sin t = \sin t$$

Then we know the velocity-time function has the equation:

$$v(t) = \cos t$$

And the acceleration-time function has the equation:

$$a(t) = -\sin t$$

(b) The velocity is zero at a time, t, when $\cos t = 0$. Thus, $v = 0$ at $t = \frac{\pi}{2}, \frac{3\pi}{2}$, etc.

(d) The acceleration is zero when $-\sin t = 0$. Thus, $a = 0$ at $t = 0, \pi, 2\pi$, etc.

2. Answers:

(a) The total momentum of the masses is equal to the momentum of the first mass, since the second mass is initially at rest. The first mass begins with potential energy equal to $PE = mgh = (5 \text{ kg})(9.8 \text{ m/s}^2)(2.5 \text{ m}) = 1{,}225$ J, and it is completely converted to kinetic energy at the bottom of the ramp.

When this energy is used to find speed, we find that $KE = 122.5 \text{ J} = \frac{1}{2}mv^2 = \frac{1}{2}5v^2$, and the speed of the mass is 7 m/s.

We can then find the initial momentum: $p = mv = (5 \text{ kg})(7 \text{ m/s}) = 35$ kg·m/s

Thus after the collision, the total momentum of the system must be 35 kg·m/s.

(b) With a momentum of 35 kg·m/s, and a combined mass of 10 kg (the two masses are stuck together), the final speed is 3.5 m/s. Substituting this value into the kinetic energy expression, we find that $KE = \left(\frac{1}{2}\right)(10)(3.5)^2 = 61.25$ J. Thus, 61.25 Joules remain, and 1,164 Joules must have been "lost" to thermal energy during the collision.

(c) To find the amplitude of the oscillation, we must know the maximum compression of the spring. We can do this by converting the 61.25 J of kinetic energy into spring potential energy: 61.25 J $= \frac{1}{2}kA^2 = \left(\frac{1}{2}\right)(200)A^2$, and $A = 0.783$ meters, or 78.3 cm.

(d) The maximum speed will occur when all of the spring potential energy has been converted back into kinetic energy. This happens exactly where the collision first occurred—what we would call the "equilibrium point." The value of the maximum speed is equal to the original speed of the two masses immediately following the collision: 3.5 m/s.

If you'd like to try this with another method, convert the spring potential energy back into kinetic energy and solve for the speed. You'll get the same answer.

(e) Maximum acceleration occurs when the force is at a maximum. We know that this occurs when the spring is completely compressed and $x = 0.783$ m. Using Hooke's law ($F = kx$), we find that the restoring force (we aren't concerned with the negative sign here) equals $F = kx = (200)(0.783) = 156.6$ N.

Knowing the force that is accelerating the masses, we use Newton's second law to find the acceleration:

$$a = \frac{F}{m} = \frac{156.6}{10} = 15.66 \text{ m/s}^2$$

3. Answers:

(a) The formula for the angles of the maxima for a diffraction grating is

$\sin \theta_m = \frac{(m\lambda)}{d}$, where d is the slit separation and $m = 0, 1, 2, \ldots$

$d = \frac{1}{L}$, where $L =$ the number of lines/meter

Thus, $d = \dfrac{1}{[(4{,}000 \text{ lines/cm})(100\text{cm/m})]} = 2.5 \times 10^{-6}$ m.

The first- and second-order maxima are for $m = 1$ and $m = 2$. For $m = 1$,

$$\sin \theta_1 = \frac{(1)(580 \times 10^{-9})}{2.5 \times 10^{-6}} = 0.232$$

$$\theta_1 = 13.4 \text{ degrees.}$$

Similarly for $m = 2$, $\theta_2 = 27.7$ degrees.

(b) X-rays have very small wavelengths (on the order of .01 nm). These wavelengths are so small that they're close to the spacing of crystals. This means that the lattices in the crystals can diffract the X-rays. The spacing in the crystals doesn't diffract the larger wavelengths of visible light.

4. Answers:

(a)

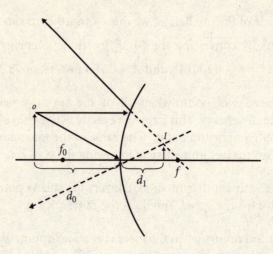

(b)

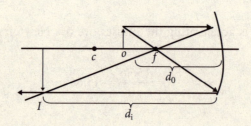

(c)

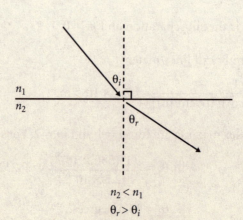

5. Answers:

(a)

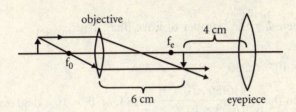

(b)

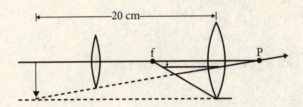

The upside-down image located 4 cm to the left of the eyepiece (right) lens is the image formed by the first lens. The image formed by the first lens is also the object for the second lens.

(c) The magnification is a combination of the first and second magnifications. The objective lens gives an image distance of 6 cm. Since the original object distance was 2 cm, the first magnification is –3.

The second (eyepiece) lens uses an object distance of 4 cm and produces an image that is at an image distance of –20 cm. Thus, the magnification is +5.

Taken together, the magnification for this microscope is (–3)(+5) = –15. This means the image is 15 times larger and inverted.

(d) A telescope looks at objects that are far enough away that the rays of light enter the telescope parallel to one another. This means the first image is formed at the focus of the objective lens, so it is **much** smaller than the original object.

6. Answer:

A dark fringe occurs when the two paths in the interferometer are out of phase by 180 degrees. When the mirror is moved by $\frac{1}{2}$ of a wavelength, the distance the light travels down and back between the mirror and the beam splitter is one wavelength. As this movement occurs, you'll see a dark fringe, then a bright fringe, then a dark fringe. So, when 200 fringes pass, the mirror has moved by this amount:

$$(200)(\frac{1}{2\lambda}) = (200)(300 \text{ nm}) = 60,000 \text{ nm} = 0.060 \text{ mm}$$

7. Answers:

(a) $\sin \theta_m = \dfrac{(m\lambda)}{d}$, where d is the slit separation, and in this case, $m = 4$.

$d = \dfrac{(0.0254 \text{ m})}{L}$, where L = the number of lines/inch. Thus,

$d = \dfrac{0.0254}{4,000} = 6.35 \times 10^{-6}$ m. So,

$$\sin \theta_4 = \dfrac{(4)(450 \times 10^{-9})}{(6.35 \times 10^{-6})} = .283, \text{ or } \theta_4 = 16.4 \text{ degrees.}$$

(b) Using $\sin \theta_m = \dfrac{(m\lambda)}{d}$ for $m = 15$ yields:

$$\sin \theta_{15} = \dfrac{(15)(450 \times 10^{-9})}{6.35 \times 10^{-6}} = 1.06,$$

which can't happen ($\sin \theta_m$ can't be greater than 1).

8. Answer:

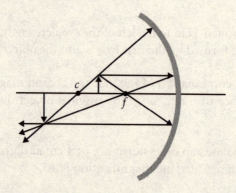

The drawing should look something like the one shown here. One can see that the magnification is approximately –2. The minus sign expresses the fact that the image is upside down.

9. Answer:

The image is formed behind the mirror and thus is virtual. The drawing should look something like this, with the image formed at the intersection of the dashed lines—upright and reduced in size:

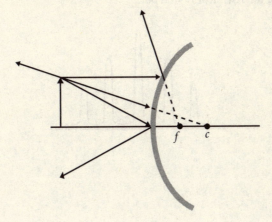

This is how approaching cars appear in rearview mirrors, so use caution!

10. Answers:

(a) *(rotated 90 degrees):*

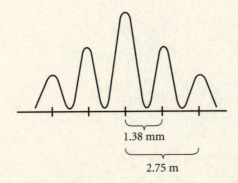

(b) The central bright fringe is found directly in a straight-line path from the original path of the light. The first maximum is found at an angle of:

$$\theta = \sin^{-1}[(1)(\frac{\lambda}{d})] = \sin^{-1}[(1)(550)(\frac{10^{-9}}{.002})] = 0.015 \text{ degrees}$$

Since the sheet is 5 meters from the slit apparatus, the distance from the first bright fringe to the center is $x = 5 \tan(.015°) = 1.38$ mm.

The second maximum from the center is found at an angle of:

$$\theta = \sin^{-1}[(2)(\frac{\lambda}{d})] = \sin^{-1}[(2)(\frac{550 \cdot 10^{-9}}{.002})] = 0.0315 \text{ degrees}$$

Thus, the distance from center is $x = 5 \tan(.0315°) = 2.75$ mm.

(c) *(rotated 90 degrees):*

With the diffraction grating, the pattern would be similar in spread, but each maximum, or bright fringe, would be narrower and sharper.

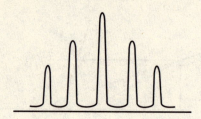

Atomic and Nuclear Physics

CHAPTER 9

I. ATOMIC AND QUANTUM PHYSICS

The *atom* is the smallest particle of an element that can be identified with that element. The atom consists of a nucleus surrounded by *electrons* which are in *quantized*, or discrete, *energy levels*. An electron can only change energy levels when it absorbs or emits energy. The energy emitted as a result of a downward energy level transition is typically in the form of a *photon*, the smallest particle of light. The photon nature of light is the principle behind the *photoelectric effect*, in which the absorption of photons of a certain frequency causes electrons to be emitted from a metal surface. The *Compton Effect* also verifies the photon nature of light. Since light waves exhibit particle (photon) properties, *de Broglie* suggested that particles, such as electrons, can exhibit wave properties.

Key Terms

alpha particle: positively charged particle consisting of two protons and two neutrons

atom: the smallest particle of an element that can be identified with that element; the atom consists of protons and neutrons in the nucleus, and electrons orbiting around the nucleus.

Compton Effect: the interaction of photons with electrons resulting in the increased wavelengths of the photons and kinetic energy of the electrons

de Broglie wavelength: the wavelength associated with a moving particle with a momentum *mv*

electron: the smallest negatively charged particle; electrons orbit the nucleus of the atom

KAPLAN 239

AP Physics

CHAPTER 9

energy level: amount of energy an electron has while in a particular orbit around the nucleus of an atom

excited state: the energy level of an electron in an atom after it has absorbed energy

ground state: the lowest energy level of an electron in an atom

Heisenberg Uncertainty Principle: the more accurately one determines the position of a subatomic particle, the less accurately its momentum is known

line spectrum: discrete lines which are emitted by a cool excited gas

photoelectric effect: the ejection of electrons from certain metals when exposed to light of a minimum frequency

photon: the smallest particle of light

Planck's Constant: the constant of proportionality between energy and frequency of a photon

quantized: the state of a quantity that cannot be divided into smaller increments forever, for which there exists a minimum, quantum increment

quantum mechanics: the study of the properties and behavior of matter using its wave properties

quantum model of the atom: atomic model in which only the probability of locating an electron is known

work function: the minimum energy required to release an electron from a metal

Quantum Phenomena

The word *quantum* simply means *the smallest piece of something*. The quantum of American money is one cent; the quantum of negative charge is the electron, since as far as we know there is no smaller negative charge that exists by itself. There are several quantities in physics which are *quantized*, that is, that occur in multiples of some smallest value. Light is one of these quantities.

Photons and the Photoelectric Effect

In prior sections we treated light as a wave. But there are circumstances when light behaves more like it is made up of individual bundles of energy, separate from each other but sharing a wavelength, frequency, and speed. The quantum of light is called the *photon*. In the late 19th century an effect was discovered by Heinrich Hertz which could not be explained by the wave model of light. He shone ultraviolet light on a piece of zinc metal, and the metal became positively charged. Although he did not know it at the time, the light was causing the metal

240 KAPLAN

to emit electrons. Using light to cause electrons to be emitted from a metal is called the *photoelectric effect*. According to the theory of light at the time, light was considered a wave, and should not be able to "knock" electrons off of a metal surface. At the turn of the 20th century, Max Planck showed that light could be treated as tiny bundles of energy called photons, and the energy of a photon was proportional to its frequency. Thus, a graph of photon energy *E* vs. frequency *f* looks like this:

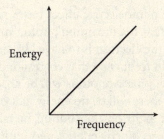

The slope of this line is a constant that occurs many times in the study of quantum phenomena called *Planck's Constant*. Its symbol is *h*, and its value is 6.67×10^{-34} J·s (or J/Hz). The equation for the energy of a photon is:

$$E = hf$$

Or, since, $f = \dfrac{c}{\lambda}$:

$$E = \dfrac{hc}{\lambda}$$

The energy of a photon is proportional to its frequency, but inversely proportional to its wavelength. This means that a violet photon has a higher frequency and energy than a red photon.

Often, when dealing with small amounts of energy like those of photons or electrons, we may prefer to use a very small unit of energy called the *electron-volt (eV)*. The conversion between joules and electron-volts is:

$$1 \ eV = 1.6 \times 10^{-19} \ J$$

Planck's Constant can be expressed in terms of electron-volts as:

$$h = 4.14 \times 10^{-15} \ eV \cdot s$$

In 1905, Albert Einstein used Planck's idea of the photon to explain the photoelectric effect: one photon is absorbed by one electron in the metal surface, giving the electron enough energy to be released from the metal. But not just any photon will knock an electron off a metal surface. The photon must first have enough energy to "dig" the electron out of the metal, and then have some energy left over to give the electron some kinetic energy to escape completely.

AP Physics
CHAPTER 9

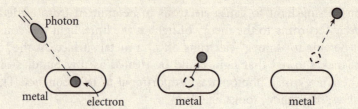

Each metal that can exhibit the photoelectric effect has a minimum energy and frequency called the *threshold frequency* f_o that the incoming photon must meet to dig the electron out of the metal, and must exceed if the electron is to have kinetic energy to escape. For example, the metal sodium has a threshold frequency that corresponds to yellow light. If yellow light is shone on a sodium surface, the yellow photons will be absorbed by electrons in the metal, causing them to be released, but there will be no energy left over for the electrons to have any kinetic energy. If we shine green light on the sodium metal, the electrons will be released, have some energy left over to use as kinetic energy, and jump off the metal completely, since green light has a higher frequency and energy than yellow light. If a brighter (more photons) green light is shone on the surface, more electrons will be emitted, since one photon can be absorbed by one electron. If these electrons are funneled into a circuit, we can use them as current in an electrical device. If orange light were shone on the sodium metal, no emission of electrons would take place, no matter how bright the orange light is, since orange light is below the threshold frequency for sodium.

The graph of maximum kinetic energy of a photoelectron vs. frequency of incident light looks like this:

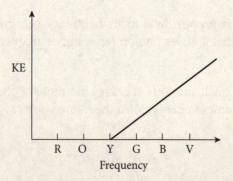

Note that the electrons have no kinetic energy up to the threshold frequency (color), and then their kinetic energy is proportional to the frequency of the incoming light.

The minimum energy needed to eject an electron from the atom completely is called the *work function* ϕ. The work function is proportional to the threshold frequency f_o by the equation

$$\phi = hf_o$$

Thus, the kinetic energy of the ejected electron from a photoemissive surface is equal to the difference between the energy of the absorbed photon and the work function:

$$K_{electron} = E_{photon} - \phi = hf - hf_o$$

Physics Review

ATOMIC AND NUCLEAR PHYSICS

The Momentum of a Photon

Since a photon has energy, does it follow that it has momentum? Recall that momentum can be defined as the product of mass and velocity. But a photon has no mass. It turns out that in quantum physics photons do have momentum by virtue of their wavelength. The equation for the momentum of a photon is:

$$p = \frac{h}{\lambda}$$

So the momentum of a photon is inversely proportional to its wavelength. Photons can, and do, impart momentum to sub-atomic particles in collisions that follow the law of conservation of momentum. This phenomenon was experimentally verified by Arthur Compton in 1922. Compton aimed x-rays of a certain frequency at electrons, and when they collided and scattered, the x-rays were measured to have a lower frequency, indicating less energy and momentum. The scattering of x-ray photons from an electron with a loss in energy of the x-ray photon is called the *Compton Effect*. It is difficult to understand how a photon, having only energy and no mass, can collide with a particle like an electron and change its momentum, but this has been verified experimentally many times.

Since a photon is the smallest and most unobtrusive measuring device we have available to us, (and even a photon has too large of a momentum to make accurate measurements of the speed and position of sub-atomic particles), we must admit to an uncertainty that will always exist in quantum measurements. This limit to accuracy at this level was formulated by Werner Heisenberg in 1928 and is called the *Heisenberg Uncertainty Principle*. It can be stated like this:

There is a limit to the accuracy of the measurement of the speed (or momentum) and position of any sub-atomic particle. The more accurately we measure the speed of a particular particle, the less accurately we can measure its position, and vice-versa.

Matter Waves and the de Broglie Wavelength of a Particle

In 1924, Louis de Broglie reasoned that if a wave such as light can behave like a particle, having momentum, then why couldn't particles behave like waves? If the momentum of a photon can be found by the equation $p = \frac{h}{\lambda}$, then the wavelength can be found by $\lambda = \frac{h}{p}$. De Broglie suggested that for a particle with mass m and speed v, we could write the equation as $\lambda = \frac{h}{mv}$, and the wavelength of a moving particle could be calculated. This hypothesis was initially met with a considerable amount of skepticism until it was shown in 1927 by Davisson and Germer that electrons passing through a nickel crystal were diffracted through the crystal, producing a diffraction pattern on a photographic plate. Thus, de Broglie's hypothesis that particles could behave like waves was experimentally verified. Nuclear and particle physicists must take into account the wave behavior of subatomic particles in their experiments. We typically don't notice the wave properties of objects moving around us

KAPLAN 243

because the masses are large in comparison to subatomic particles and the value for Planck's constant h is extremely small. But the wavelength of any moving mass is inversely proportional to the momentum of the object.

Atomic Physics

The ancient Greeks were the first to document the concept of the atom. They believed that all matter was made up of tiny indivisible particles. In fact, the word *atom* comes from the Greek word *atomos*, meaning "uncuttable." But a working model of the atom didn't begin to take shape until J.J. Thomson's discovery of the electron in 1897. He found that electrons are tiny negatively charged particles and that all atoms contain electrons. He also recognized that atoms are naturally neutral, containing equal amounts of positive and negative charge, although he was not correct in his theory of how the charge was arranged.

You may remember studying Thomson's "plum-pudding" model of the atom, with electrons floating around in positive fluid. A significant improvement on this model of the atom was made by Ernest Rutherford around 1911, when he decided to shoot alpha particles (helium nuclei) at very thin gold foil to probe the inner structure of the atom. He discovered that the atom has a dense, positively charged nucleus with electrons orbiting around it. In 1913, Niels Bohr made an important improvement to the Rutherford model of the atom. He observed that excited hydrogen gas gave off a spectrum of colors when viewed through a spectrosope. But the spectrum was not continuous; the colors were bright, sharp lines that were separate from each other. It had long been known that every low pressure, excited gas emitted its own special spectrum in this way, but Bohr was the first to associate the bright-line spectra of these gases, particularly hydrogen, with a model of the atom.

He proposed that the electrons orbiting the nucleus of an atom do not radiate energy in the form of light while they are *in* a particular orbit, but only when they *change* orbits. Furthermore, an electron cannot orbit at just any radius around the nucleus, but only certain selected (quantized) orbits.

The two postulates of the Bohr model of the atom are summarized below:

1. Electrons orbiting the nucleus of an atom can only orbit in certain quantized orbits, and no others. These orbits from the nucleus outward are designated $n = 1, 2, 3\ldots$, and the electron has energy in each of these orbits E_1, E_2, E_3, and so on. The energies of electrons are typically measured in *electron-volts (eV)*. The lowest energy (in the orbit nearest the nucleus) is called the *ground state energy*.

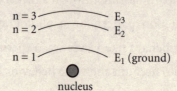

2. Electrons can change orbits when they absorb or emit energy.

   (a) When an electron absorbs *exactly* enough energy to reach a higher energy level, it jumps up to that level. If the energy offered to the electron is not *exactly* enough to raise it to a higher level, the electron will ignore the energy and let it pass.

   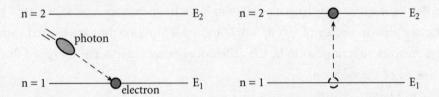

   (b) When an electron is in a higher energy level, it can jump down to a lower energy level by releasing energy in the form of a photon of light. The energy of the emitted photon is exactly equal to the difference between the energy levels between which the electron moves.

   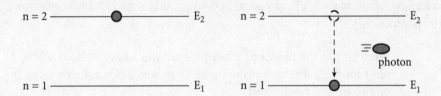

For example, consider the energy level diagram for a particular atom shown below:

Energy, E (eV)	Energy state, n
−0.38	4
−0.85	3
−3.4	2
−13.6	1

Let's say an electron in the ground state absorbs a photon and makes an upward transition to $n = 4$. Since the energy levels are labeled with the energy of the electron above ground state in each case, the electron would need 13.2 eV to jump from the ground state energy to $n = 4$.

If the electron drops from $n = 4$ to $n = 2$, a photon of energy 3 eV is emitted. If the electron then drops from E_2 to E_1, a photon of energy 3.02 eV is emitted. The second transition results in a higher energy photon than the first, and since energy is proportional to frequency the second photon must also have a higher frequency.

AP Physics

CHAPTER 9

Sample Problem:

Monochromatic red light ($\lambda = 700$ nm) is shone onto a metal surface that ejects electrons with a kinetic energy of 1.1 eV.

(a) What is the work function of the metal?

Answer: Red light with a wavelength of 700 nm has a frequency of $\frac{c}{\lambda} = 4.29 \times 10^{14}$ Hz. This translates to a photon energy of $E = hf = 1.77$ eV. Since the electrons are ejected with 1.1 eV of energy, the work function has to be the difference between the initial energy of the photon and the energy of the electron.

$$1.77 \text{ eV} - 1.1 \text{ eV} = 0.67 \text{ eV}$$

(b) If a typical fission reaction releases 170 MeV, then how many photons of the red light would be required to produce electrons with a combined energy equal to the energy of a single fission reaction?

Answer: Each electron has 1.1 eV of energy. Dividing this into 170 MeV gives us a total of 1.54×10^8 photons.

(c) What would the power of a beam of red light have to be if all the photons were directed at the metal to produce this much energy in one second of operation?

Answer: To generate this much energy, there would have to be 1.54×10^8 red light photons delivered each second. This is equivalent to 1.77 eV times 1.54×10^8 photons, which is 2.74×10^8 eV. This equals 4.38×10^{-11} J per second, or 4.38×10^{-11} watts.

(d) List two changes you could make to the light to accomplish the same effect (matching the energy of a single fission reaction) more quickly.

Answer: If you increased the intensity of the light (made it brighter), you would produce the necessary energy faster, as more electrons would be produced. If you could increase the frequency of the light (make it more "blue"), then there would be the same number of electrons would be produced, but each electron would have more energy, as each incident photon would have more energy and the work function stays the same.

II. NUCLEAR PHYSICS

The modern view of the atom includes electrons in energy levels around the *nucleus* of the atom. The nucleus contains positively charged *protons* and neutral *neutrons*, each of which are made up of *quarks*. The nucleus is held together by the *strong nuclear force*, and the *binding energy* in the nucleus is a result of some of the mass of the particles (the *mass defect*) in the nucleus being converted into energy by the relationship $E = mc^2$. Nuclear changes can take place, but the total amount of atomic mass in the process must remain constant.

246   **KAPLAN**

Key Terms

alpha particle: positively charged particle consisting of two protons and two neutrons

atomic mass unit: the unit of mass equal to 1/12 the mass of a carbon-12 nucleus

atomic number: the number of protons in the nucleus of an atom

beta particle: high speed electron emitted from a radioactive element when a neutron decays into a proton

binding energy: the nuclear energy that binds protons and neutrons in the nucleus of the atom

chain reaction: nuclear process producing more neutrons which in turn can create more nuclear processes; usually applied to fission

critical mass: the minimum amount of mass of fissionable material necessary to sustain a nuclear chain reaction

element: a substance made of only one kind of atom

fundamental particles: the particles (quarks and leptons) of which all matter is composed

isotope: a form of an element which has a particular number of neutrons, that is, it has the same atomic number but a different mass number than the other elements which occupy the same place on the periodic table

mass defect: the mass equivalent of the binding energy in the nucleus of an atom by $E = mc^2$

neutron: an electrically neutral subatomic particle found in the nucleus of an atom

nuclear fission: the splitting of a heavy nucleus into two smaller ones

nuclear fusion: the combining of two light nuclei into one larger one

nuclear reaction: any process in the nucleus of an atom that causes the number of protons and/or neutrons to change

nuclear reactor: device in which nuclear fission or fusion is used to generate electricity

nucleon: either a proton or a neutron

quark: one of the elementary particles of which all protons and neutrons are made

strong nuclear force: the force that binds protons and neutrons together in the nucleus of an atom

AP Physics

CHAPTER 9

Nuclear Structure

The nucleus is made up of positively charged protons and neutrons, which have no charge. The proton has exactly the same charge as an electron, but is positive. The neutron is actually made up of a proton and an electron bound together to create the neutral particle. A proton is about 1,800 times more massive than an electron, which makes a neutron only very slightly more massive than a proton. We say that a proton has a mass of approximately one *atomic mass unit, u*. The *atomic number (Z)* of an element is equal to the number of protons found in an atom of that element, and fundamentally is an indication of the charge on the nucleus of that element. All atoms of a given element have the same atomic number. In other words, the number of protons an atom has defines what kind of element it is. The total number of neutrons and protons in an atom is called the *mass number (A)* of that element. The symbol $^A_Z X$ is used to show both the atomic number and the mass number of an X atom, where Z is the atomic number and A is the mass number.

Even though the number of protons must be the same for all atoms of a particular element, the number of neutrons, and thus the mass number, can be different. Atoms of the same element with different masses are known as *isotopes* of one another. For example, carbon-12 is a carbon atom with 6 protons and 6 neutrons, while carbon-14 is a carbon atom with 6 protons and 8 neutrons. We would write these two isotopes of carbon as $^{12}_6 C$ and $^{14}_6 C$.

The table below summarizes the basic features of protons, neutrons and electrons. Notice that we use an H to symbolize the proton, since the proton is a hydrogen nucleus.

Particle	Symbol	Relative mass	Charge	Location
proton	$^1_1 H$	1	+1	nucleus
neutron	$^1_0 n$	1	0	nucleus
electron	$^{\ 0}_{-1} e$ or e^-	0	−1	electron orbitals around the nucleus

For example, consider a neutral atom of iron $^{56}_{26} Fe$. This isotope of iron has an atomic number of 26 and a mass number of 56. Therefore, it will have 26 protons, 26 electrons, and $56 - 26 = 30$ neutrons.

Nuclear Binding Energy

Since positive charges repel each other, one might wonder why protons would want to stay together in the nucleus of the atom. There must be a force holding the protons together which is greater than the electrostatic repulsion between them. This force is called the strong nuclear force, and is a result of the binding energy of the nucleus. But where does this energy which holds the nucleus together come from?

248 **KAPLAN**

Physics Review

ATOMIC AND NUCLEAR PHYSICS

According to Einstein's famous equation $E = mc^2$, mass and energy can be considered to be interchangeable. As the nucleus is assembled, each proton and neutron gives up a little of its mass to be converted into binding energy. For example, if you start with two protons and two neutrons, you have a total of 4 atomic mass units (u). But if these particles are combined into a helium $^4_2$He nucleus, the resulting mass of the helium nucleus is less than 4 atomic mass units, since some of the mass of the protons and neutrons has been converted into binding energy to hold the nucleus together. Likewise, when a nucleus is split, it doesn't need all of its original binding energy anymore, and some of that energy is released as heat. The equivalence between mass in atomic mass units and energy in million electron volts (MeV) is:

$$1 \text{ u} = 931 \text{ } MeV$$

This conversion factor, along with the masses of electrons, protons, and neutrons is listed in the table of information provided on the AP Physics B exam.

Nuclear Reactions

At the end of the 19th century, there were elements discovered that continuously emitted mysterious rays. These elements were identified as being *radioactive*. A radioactive element spontaneously emits particles from its nucleus because the energy of the nucleus is unstable. Examples of naturally-occurring radioactive elements are uranium $^{238}_{92}$U, radium $^{226}_{88}$Ra, and carbon $^{14}_{6}$C.

There are four types of particles that can be emitted when an element undergoes radioactive decay:

1. *Alpha decay.* Uranium, for example, undergoes *alpha decay*, meaning that it emits an *alpha particle* from its nucleus. An alpha particle is a helium nucleus, consisting of 2 protons and 2 neutrons. When an element emits an alpha particle, its nucleus loses 2 atomic numbers and 4 mass numbers, and thus changes into another element, called the *daughter element*. But what would this element be? We can write the nuclear equation for the radioactive decay of uranium as

$$^{238}_{92}U \rightarrow {}^A_Z X + {}^4_2 He$$

    where X is the daughter element and $^4_2$He is the alpha particle. The atomic number on the left must equal the sum of the atomic numbers on the right, since charge and mass are conserved in this process. The same is true for the mass numbers on the left and right. So, the daughter element has an atomic number Z = 92 − 2 = 90 and a mass number A = 238 − 4 = 234. Uranium decays into the daughter element $^{234}_{90}$Th, thorium.

2. *Beta decay.* A *beta particle* is the name given to an electron emitted from the nucleus of a radioactive element. But what is an electron doing in the nucleus of an atom? Remember

KAPLAN   249

that we discussed the neutron in the nucleus of an atom as being a proton and an electron bound together. Beta decay is really just a neutron emitting an electron and becoming a proton. Thus, the daughter element resulting from beta decay is one atomic number higher than the parent nucleus, but the mass number essentially does not change. For example, carbon $^{14}_{6}C$ is a radioactive element that undergoes beta decay. The decay equation is:

$$^{14}_{6}C \rightarrow {}^{A}_{Z}X + {}^{0}_{-1}e$$

We use the same symbol for a beta particle as we do for an electron. The daughter element must have an atomic number of $6 - (-1) = 7$ and a mass number of $14 - 0 = 14$. The daughter element is $^{14}_{7}N$, nitrogen.

3. *Gamma decay.* Some radioactive elements emit a gamma ray, a very high energy electromagnetic wave which has no charge or mass, so only the energy of the nucleus changes, and neither Z nor A change.

4. *Positron decay.* A positron is exactly like an electron except for the fact that it is positively charged. A positron is not a proton, as their masses and other features are very different. Positron decay equations are typically not included on the AP Physics B exam.

Fusion

Fusion occurs when small nuclei combine into larger nuclei and energy is released. Many stars, including our sun, power themselves by fusing four hydrogen nuclei to make one helium nucleus. The fusion process is clean and a large amount of energy is released, which is why researchers here on Earth are trying to find ways to use fusion as an alternative energy source.

For example, the element tritium $^{3}_{1}H$ is combined with $^{2}_{1}H$, a hydrogen isotope called deuterium, to form helium $^{4}_{2}He$ and a neutron, along with the release of energy. The equation for this fusion reaction is:

$$^{3}_{1}H + {}^{2}_{1}H \rightarrow {}^{4}_{2}He + {}^{1}_{0}n$$

Note that the sum of the atomic numbers on the left must equal the sum of the atomic numbers on the right. The same is true of the mass numbers.

Fission

Fission is a process in which a large nucleus splits into smaller nuclei. Fission is usually caused artificially by shooting a slow neutron at a large atom, such as uranium, which absorbs the neutron, becomes unstable, and splits into two smaller atoms, along with the release of more neutrons and some energy.

Physics Review

ATOMIC AND NUCLEAR PHYSICS

For example, a fission reaction occurs when uranium $^{235}_{92}U$ absorbs a slow neutron and then splits into barium and krypton, releasing three neutrons and some energy. The equation for this fission reaction is:

$$^{235}_{92}U + ^{1}_{0}n \rightarrow ^{141}_{56}Ba + ^{96}_{36}Kr + 3^{1}_{0}n + energy$$

Once again, the sum of the atomic and mass numbers on the left equal the sum of the atomic and mass numbers on the right. The three neutrons which are produced from this reaction can be used to split three more uranium atoms, which produce three more neutrons in each reaction, each of which can split three more uranium atoms, and so on. This is called a *chain reaction*, and is used to sustain the release of energy in fission reactions. However, before a chain reaction can be sustained, there must be a minimum amount of fissionable material, such as uranium or plutonium, present. The minimum amount of fissionable material that must be present to sustain a chain reaction is called the *critical mass*.

Sample Problem:

The fission reaction $n + ^{235}U \rightarrow ^{236}U^{*} \rightarrow ^{141}Ba + ^{92}Kr + 3n$ produces 170 MeV of kinetic energy.

(a)   How many of these fission events are needed to produce energy of 1 kilowatt-hour (kWh), that is, the energy it takes to run your blow dryer for an hour?
(Helpful unit conversion: $1 \text{ eV} = 1.602 \times 10^{-19}$ J)

Answer:

Convert the energy needed to Joules:

$$1 \text{ kWh} = 3.6 \text{ MJ}$$

Then convert kinetic energy produced to Joules:

$$170 \text{ MeV} = 1.7 \times 10^{8} \text{ eV} = (1.7 \times 10^{8} \text{ eV})(1.6 \times 10^{-19} \text{ J/eV}) = 2.7 \times 10^{-11} \text{ J}$$

Find the number of fission events, N:

$$(N)(170 \text{ MeV}) = 1 \text{ kWh}$$

$$N = \frac{3.6 \times 10^{6} \text{ J}}{2.7 \times 10^{-11} \text{ J}} = 1.3 \times 10^{17} \text{ events}$$

(b)   How many neutrons are produced in this chain reaction process?

Answer:

Each reaction consumes one neutron and produces three, for a net gain of two. So, a total of 2.6×10^{17} neutrons are produced.

KAPLAN   251

AP Physics
CHAPTER 9

(c) How much does this amount of Uranium weigh? (The atomic mass for Uranium-235 is 235.0439 u; one atomic mass unit is 931.5 MeV/c^2 or 1.6605×10^{-27} kg.)

Answer:

The mass of ^{235}U is $(235.0439 \text{ u})(1.6605 \times 10^{-27} \text{ kg/u}) = 3.903 \times 10^{-25}$ kg.

The total mass is simply given by 1.3×10^{17} times the mass of a single uranium atom:

$$M = (1.3 \times 10^{17})(3.903 \times 10^{-25} \text{ kg}) = 5.1 \times 10^{-8} \text{ kg} = 0.051 \ \mu g$$

Important Equations

Energy of a wave related to frequency and Planck's Constant:	$E = hf$
Energy of a wave related to momentum:	$E = pc$
Equation for photoelectric effect:	$K_{max} = hf - \Phi$
De Broglie wavelength:	$\lambda = \dfrac{h}{p}$
Mass-energy equivalence:	$\Delta E = (\Delta m)c^2$

ATOMIC AND NUCLEAR PHYSICS

PRACTICE PROBLEMS

1. The atomic masses of the hydrogen isotopes are:

    hydrogen = $_1^1H$: 1.007825 u

    deuterium = $_1^2H$: 2.014102 u

    tritium = $_1^3H$: 3.016049 u

    (a) What is the energy released in the reaction $_1^2H + _1^2H \rightarrow _1^3H + _1^1H$? Give your answer in units of MeV.//
    (b) What mass of deuterium ($_1^2H$) would be needed to generate one kWh?//
    (*Hint: one atomic mass unit is $u = 1.6605 \times 10^{-27}$ kg.*)

2. You start with 1,090 μg of a material with a half-life of 80 seconds.
    (a) How many μg are left after 160 seconds?
    (b) Draw a diagram of the mass of the material as a function of time.

3. What is the approximate half-life of the radioactive isotope for which the activity is plotted below?

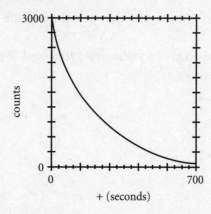

4. Write the reaction equations with real nuclei for the four different radioactive decays below.
    (a) Alpha decay of $_{95}^{241}Am$
    (b) Beta- decay of $_6^{14}C$
    (c) Beta+ decay of $_9^{18}F$
    (d) Gamma decay of $_{66}^{152}Dy$

AP Physics
CHAPTER 9

5. A fusion reaction is $_1^2H + _1^3H \rightarrow _2^4He + _0^1n$, where the masses are deuterium (2.01412 u), tritium (3.01605 u), helium (4.0026 u), and neutron (1.008665 u).

(a) How much energy is released in each reaction?

(b) One gallon of gasoline produces, on average, 2.1×10^9 J of energy and costs \$1.80 per gallon. Approximately how much money could you save by using 2.01412 mg of deuterium and 3.01605 mg of tritium (assume the hydrogen costs you nothing) to burn as fuel instead of gasoline?
(*Hint: You need to find out how many moles are 2.01412 mg of deuterium and 3.01605 mg of tritium.*)

6. You set up an experiment to measure the decay of $_{84}^{212}$Polonium. The values shown below indicate the number of nuclei remaining and the time.

Number	500	225	140	61	26	10	3
Time (min)	0	1	2	3	4	5	6

(a) Make a graph of number of nuclei versus time.

(b) Using the curve from part *a*, what is the half-life of ^{212}Polonium?

(c) $_{84}^{212}$Polonium decays by alpha decay into lead. Write the equation for this decay process.

254 **KAPLAN**

ATOMIC AND NUCLEAR PHYSICS

7. An experiment performed in the physics laboratory is designed to examine the photoelectric effect. A light emitting diode (LED) is connected to a power supply, and the voltage supplied to the LED is adjusted to the point where the light is just on the verge of lighting. Several trials are run, with multiple trials on each of several different colors of LED's. The wavelength given for each color is the wavelength for peak emission as indicated by the manufacturer of the LED. Use the data below to answer the following questions.

Color	Wavelength (nm)	Voltage (v)	Frequency (Hz)	Energy (J)
red	635	1.54		
red	635	1.56		
yellow	600	1.65		
yellow	600	1.66		
green	565	1.80		
green	565	1.73		

(a) Complete the table with the frequencies for each of the trials.
(b) Assuming the wavelength reported in each trial to be the threshold wavelength, complete the table with the energy supplied by the power supply to the individual electrons on the verge of moving across the energy gap between the two terminals in the diode.
(c) On the axes below, plot "Energy (in Joules) vs. Frequency (in Hz)" and draw a best fit curve (or line).

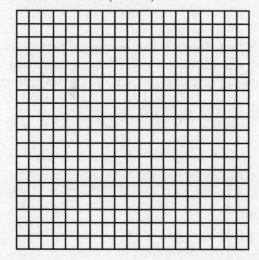

(d) Use your plot to determine a value for Planck's Constant. Explain or show clearly how you determined the value.
(e) What is the percentage error for your value of Planck's Constant?
(f) Discuss one factor in the experiment that might account for this error in the determination of Planck's Constant.

CHAPTER 9

Solutions to Practice Problems

1. Answers:

(a) The mass energy on the left-hand side is 2 × 2.014102 u = 4.028204 u.

The mass energy on the right-hand side is 3.016049 u + 1.007825 u = 4.023874 u.

Take the difference: 2 × 2.014102 u − (3.016049 u + 1.007825 u) = 0.004330 u.

Convert this to MeV: (0.004330 u)(931.5 MeV/1 u) = 4.033 MeV

(b) Each fusion reaction of two deuterium isotopes produces 4.033 MeV (see part *a*). Convert this to J:

$$4.033 \text{ MeV} = (4.033 \text{ MeV})(1.6 \times 10^{-19} \text{ J/eV}) = 6.461 \times 10^{-13} \text{ J}$$

1 kWh = 3.6 × 10^6 J.

So, we need $N = \dfrac{(3.6 \times 10^6 \text{ J})}{(6.461 \times 10^{-19} \text{ J})} = 5.57 \times 10^{18}$ fusion events.

The two deuterium isotopes needed for each fusion reaction have a combined mass of:

$$M = 2 \times 2.014102 \text{ u} = 4.028204 \text{ u} = 4.028204 \text{ u} \times (1.6605 \times 10^{-27} \text{ kg/1 u}) = 6.6888 \times 10^{-27} \text{ kg}$$

Multiply this by N and get the total mass required:

$$M_{total} = (5.57 \times 10^{18})(6.6888 \times 10^{-27} \text{ kg}) = 3.7 \times 10^{-8} \text{ kg}$$

2. Answers:

(a) The original mass will be reduced by half during the first 80 seconds, which is one half life. During the second 80 seconds, the mass will be cut in half again, so that only one-fourth of the original mass remains. Thus, after 160 seconds, the amount remaining is:

$$(0.25)(1{,}090 \text{ μg}) = 272.5 \text{ μg}$$

(b)

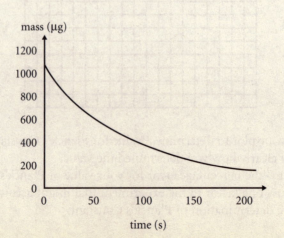

3. Answer:

The half-life is the time it takes for $\frac{1}{2}$ of the original nuclei to decay.

(Remember: $t_{1/2} = 0.693\ \tau$). So we just look in the graph to determine the time at which only half of the original 3000 nuclei remain:

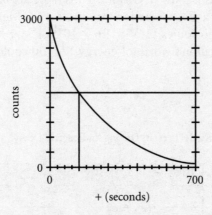

The horizontal line across the middle represents the line at which only 1,500 nuclei are left. The vertical line indicates the time at which this happens. Since the full scale is 700 s, each of the time divisions is $\frac{700}{5}$ s = 140 s. So the half-life is slightly larger than 140 s, perhaps 150 seconds.

4. Answers:

(a) Alpha decay of $^{241}_{95}\text{Am}$:

$^{241}_{95}\text{Am} \rightarrow\ ^{237}_{93}\text{Np} +\ ^{4}_{2}\text{He}$ or $^{241}_{95}\text{Am} \rightarrow\ ^{237}_{93}\text{Np} + \alpha$

(b) Beta decay of $^{14}_{6}\text{C}$:

$^{14}_{6}\text{C} \rightarrow\ ^{14}_{7}\text{N} + \nu_e + e^-$ (or $^{14}_{6}\text{C} \rightarrow\ ^{14}_{7}\text{N} +\ ^{0}_{-1}e$)

(c) Beta+ decay of $^{18}_{9}\text{F}$:

$^{18}_{9}\text{F} \rightarrow\ ^{18}_{8}\text{O} + \nu_e + e^+$ (or $^{18}_{9}\text{F} \rightarrow\ ^{18}_{8}\text{O} +\ ^{0}_{+1}e$)

(d) Gamma decay of $^{152}_{66}\text{Dy}$:

$^{152}_{66}\text{Dy} \rightarrow\ ^{152}_{66}\text{Dy} + \gamma$

5. Answers:

(a) The masses add as follows: 2.01412 u + 3.01605 u = 4.0026 u + 1.008665 u + energy. The mass equivalent of the energy equals 0.0189 u, which translates to 17.6 MeV.

(b) You need to figure out how much energy you can produce in your fusion engine. Two mg of deuterium is 0.001 mole as is 3 mg of tritium. Thus there are $0.001 \times 6.02 \times 10^{23}$ reactions taking place, or 6.02×10^{20}. Each reaction produces 17.6 MeV of energy, which equals 1.06×10^{28} eV. Converting to Joules (1eV = 1.6×10^{-19} J), we have 1.69×10^{9} J of energy produced. This is .807 of a gallon's worth of energy. So, you could save $1.45.

6. Answers:
(a)

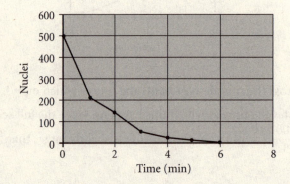

(b) The curve suggests a half-life of anywhere between 40 and 60 seconds, with a value centered around 50 seconds (0.85 minutes).

(c) $^{212}_{84}\text{Po} \rightarrow {}^{208}_{82}\text{Pb} + {}^{4}_{2}\text{He}$

7. Answers:

Color	Wavelength (nm)	Voltage (v)	Frequency (Hz)	Energy (J)
red	635	1.54	4.72441E+14	2.464E-19
red	635	1.56	4.72441E+14	2.496E-19
yellow	600	1.65	5E+14	2.64E-19
yellow	600	1.66	5E+14	2.656E-19
green	565	1.80	5.30973E+14	2.88E-19
green	565	1.73	5.30973E+14	2.768E-19

(a) To calculate the values for frequency (Hz) given the values for wavelength (nm), use the equation: $c = f\lambda$ or $f = \dfrac{c}{\lambda} = \dfrac{(3 \times 10^8 \text{ m/s})}{\lambda}$. [See calculated values in table.]

(b) To calculate the values for energy (E) in Joules given the voltages, use the equation: $eV = E$, where V is the given voltage in case and e is the charge on each electron, which is 1.6×10^{-19} C. With charge in Coulombs and voltage in volts, the energy will be in Joules. (Remember: 1 v = 1 J/C.)

(c) For this plot, be cautious of units on the axes, as they will be essential for calculations. Once you determine that the best fit for the data is a line and not a curve, draw a straight line that has data points distributed equally on both sides, but does not necessarily go through the data points.

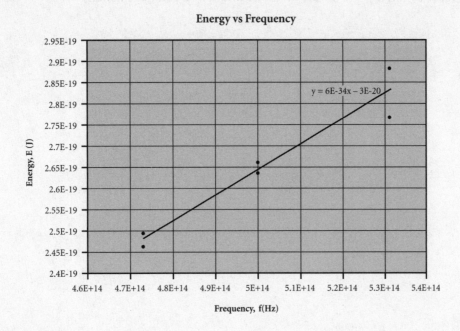

(d) The energy supplied to the electrons in the LED will be equal to the energy in the photons of light produced, so we equate the two expressions:

$$E = eV = hf$$

Since we have plotted E *vs.* f, this is of the form $y = mx$, where m is the slope. In the case above, the slope of the line is Planck's Constant, h. The slope should be determined to be about 6×10^{-34} J/Hz, which is 6×10^{-34} J·s.

(e) The accepted value for Planck's Constant is 6.2×10^{-34} J·s. The formula to calculate percentage error is:

$$\% \text{ error} = \dfrac{\text{(absolute value of error)}}{\text{accepted value}}$$

AP Physics

CHAPTER 9

$$\% \text{ error} = \frac{(0.2 \times 10^{-34})}{(6.2 \times 10^{-34})} = 0.032 \text{ or } 3.2\%$$

(f) The obvious suggestion here to improve the experiment is to run a larger number of trials, gathering more data points to use in constructing a better best fit line. One might also look at the fact that the experimental value is lower than the accepted value and look for the effect of data that might lower the experimental value. For example, the large separation of values on the green LED might indicate that to be an unreliable set of data; thus, more trials on the green would be a better indicator of where the average is for the frequency of green light. Another factor would be the difficulty of selecting the exact point to read the voltage. Unless the readings are taken in a very dark room, it would be difficult to determine when the LED is just on the verge of lighting. Reading the voltage too low would also cause a low calculation of the value of h. (Any one of these reasons would be a suitable answer.)

PRACTICE TESTS

Section III

HOW TO TAKE THE PRACTICE TESTS

This section contains two complete sample tests, each consisting of a 90-minute multiple-choice section and a 90-minute free-response section. For maximum benefit, each test should be taken in the time allotted, using only the materials allowed for that section. A sheet of fundamental constants is supplied on all sections of the AP Physics B Exam, but calculators and formula sheets are allowed only on free-response sections.

After taking each test, score your answers using the answer keys and the scoring outline at the end of the tests. This will give you an idea not only of how quickly you must work on the exam but where your time is best spent.

As you work on the multiple-choice sections of the tests, remember that there is a scoring penalty for wrong answers, so unless you know the answer, it is generally best not to guess unless you can narrow the choices to two. Remember that on the actual test, you will be marking your answers with a pencil on a machine-scored answer sheet.

As you work on the free-response sections, remember that scoring is based upon work shown in the space provided. On each problem, clearly show the formulas used with substitution of values and units. There is no penalty for substitution of an incorrect value obtained from a previous section. Thus, if a value is needed from a previous section that you were not able to work, you might define a value to substitute by simply stating, for example, "assume that the answer to part (a) is 2 m/s". Then use that value in your substitution and go ahead with solution of the problem. *Keep in mind that the readers who score the free response section of the AP Physics B Exam are looking for indications that you understand the physics of the problem.* Communicate what you know to the readers by showing your work clearly and explaining your reasoning in brief but clear statements. No extra credit is given to lengthy answers, and a student will sometimes lose credit by saying too much and contradicting a previous correct statement. Even though duplicate free-response questions are provided in a pink booklet and a green one, the pink copy is the only one that is sent in for scoring. Since time is limited, do all your work on the pink copy and use the green copy only for reference as you work. Generally speaking, you want to spend the same number of minutes on each problem as the point value of the problem.

PRACTICE TEST I HELPFUL INFORMATION (BOTH TEST SECTIONS)

Acceleration due to gravity
at the Earth's surface: $g = 9.8 \text{ m/s}^2$

Avogadro's number: $N_0 = 6.02 \times 10^{23} \text{mol}^{-1}$

Boltzmann's constant: $k_B = 1.38 \times 10^{-23} \text{ J/K}$

Coulomb's law constant: $k = 1/4\pi\varepsilon_0 = 9.0 \times 10^9 \text{ N} \cdot \text{m}^2/\text{C}^2$

Electron mass: $m_e = 9.11 \times 10^{-31} \text{ kg}$

Magnetic constant: $k' = \mu_0/4\pi = 10^{-7} \text{ (T} \cdot \text{m)/A}$

Magnitude of the electron charge: $e = 1.60 \times 10^{-19} \text{ C}$

Neutron mass: $m_n = 1.67 \times 10^{-27} \text{ kg}$

Planck's constant: $h = 6.63 \times 10^{-34} \text{ J} \cdot \text{s}$
$= 4.14 \times 10^{-15} \text{ eV} \cdot \text{s}$
$hc = 1.99 \times 10^{-25} \text{ J} \cdot \text{m}$
$= 1.24 \times 10^3 \text{ eV} \cdot \text{nm}$

Proton mass: $m_p = 1.67 \times 10^{-27} \text{ kg}$

Speed of light: $c = 3.00 \times 10^8 \text{ m/s}$

Universal gas constant: $R = 8.31 \text{ J/(mol} \cdot \text{K)}$

Universal gravitational constant: $G = 6.67 \times 10^{-11} \text{ m}^3/\text{kg} \cdot \text{s}^2$

Vacuum permeability: $\mu_0 = 4\pi \times 10^{-7} \text{ (T} \cdot \text{m)/A}$

Vacuum permittivity: $\varepsilon_0 = 8.85 \times 10^{-12} \text{ C}^2/\text{N} \cdot \text{m}^2$

1 atmosphere pressure: $1 \text{ atm} = 1.0 \times 10^5 \text{ N/m}^2$
$= 1.0 \times 10^5 \text{ Pa}$

1 electron volt: $1 \text{ eV} = 1.60 \times 10^{-19} \text{ J}$

1 unified atomic mass unit: $1 \text{ u} = 1.66 \times 10^{-27} \text{ kg}$
$= 931 \text{ MeV}/c^2$

Full-Length Practice Tests

PRACTICE TEST I HELPFUL INFORMATION (BOTH TEST SECTIONS)

UNITS

ampere	A	kilogram	kg
coulomb	C	meter	m
degree Celsius	°C	mole	mol
electron-volt	eV	newton	N
farad	F	ohm	Ω
henry	H	pascal	Pa
hertz	Hz	second	s
joule	J	tesla	T
kelvin	K	volt	V
		watt	W

PREFIXES

centi	→	10^{-2}	→	c
giga	→	10^{9}	→	G
kilo	→	10^{3}	→	k
mega	→	10^{6}	→	M
micro	→	10^{-6}	→	μ
milli	→	10^{-3}	→	m
nano	→	10^{-9}	→	n
pico	→	10^{-12}	→	p

COMMON ANGLES

θ	$\sin \theta$	$\cos \theta$	$\tan \theta$
0°	0	1	0
30°	1/2	$\sqrt{3}/2$	$\sqrt{3}/3$
37°	3/5	4/5	3/4
45°	$\sqrt{2}/2$	$\sqrt{2}/2$	1
53°	4/5	3/5	4/3
60°	$\sqrt{3}/2$	1/2	$\sqrt{3}$
90°	1	0	∞

KAPLAN 263

AP Physics

PRACTICE TEST I HELPFUL INFORMATION (FREE RESPONSE SECTION ONLY)

ATOMIC AND NUCLEAR PHYSICS

$$K_{max} = hf - \o$$

$$\Delta E = (\Delta m)c^2$$

$$E = hf = pc$$

$$\lambda = \frac{h}{p}$$

f = frequency
m = mass
λ = wavelength
E = energy
K = kinetic energy
p = momentum
$\o$ = work function

FLUID MECHANICS AND THERMAL PHYSICS

$$F_{buoy} = pVg$$

$$p + pgy + \frac{1}{2}pv^2 = \text{const.}$$

$$Q = mL$$

$$pV = nRT$$

$$v_{rms} = \sqrt{\frac{3RT}{M}} = \sqrt{\frac{3k_B T}{\mu}}$$

$$Q = nc\Delta T$$

$$e = \left| \frac{W}{Q_H} \right|$$

$$p = p_0 + pgh$$

$$A_1 v_1 = A_2 v_2$$

$$\Delta \ell = \alpha \ell_0 \Delta T$$

$$p = \frac{F}{A}$$

$$K_{avg} = \frac{3}{2} k_B T$$

$$W = -p\Delta V$$

$$\Delta U = Q + W$$

$$\Delta U = nc_v \Delta T$$

$$e_c = \frac{T_H - T_C}{T_H}$$

c = specific heat or molar specific heat
F = force
K_{avg} = average molecular kinetic energy
ℓ = length
m = mass of sample
p = pressure
T = temperature
V = volume
v_{rms} = root-mean-square velocity
y = height
μ = mass of molecule
A = area
e = efficiency
h = depth
L = heat of transformation
M = molecular mass
n = number of moles
Q = heat transferred to a system
U = internal energy
v = velocity or speed
W = work done on a system
α = coefficient of linear expansion
ρ = density

KAPLAN

Full-Length Practice Tests

PRACTICE TEST I HELPFUL INFORMATION (FREE RESPONSE SECTION ONLY)

ELECTRICITY AND MAGNETISM

$$\mathbf{E} = \frac{\mathbf{F}}{q}$$

$$E_{avg} = -\frac{V}{d}$$

$$C = \frac{Q}{V}$$

$$U_C = \frac{1}{2}QV = \frac{1}{2}CV^2$$

$$R = \frac{\rho\ell}{A}$$

$$P = IV$$

$$\frac{1}{C_s} = \sum_i \frac{1}{C_i}$$

$$\frac{1}{R_P} = \sum_i \frac{1}{R_i}$$

$$F_B = BI\ell \sin\theta$$

$$F_B = qvB \sin\theta$$

$$\Phi_m = \mathbf{B} \cdot \mathbf{A} = BA \cos\theta$$

$$\xi = B\ell v$$

$$F = \frac{1}{4\pi\varepsilon_0}\frac{q_1 q_2}{r^2}$$

$$U_E = qV = \frac{1}{4\pi\varepsilon_0}\frac{q_1 q_2}{r}$$

$$V = \frac{1}{4\pi\varepsilon_0}\sum_i \frac{q_i}{r_i}$$

$$C = \frac{\varepsilon_0 A}{d}$$

$$I_{avg} = \frac{\Delta Q}{\Delta t}$$

$$V = IR$$

$$C_P = \sum_i C_i$$

$$R_s = \sum_i R_i$$

$$B = \frac{\mu_0}{2\pi}\frac{I}{r}$$

$$\xi_{avg} = -\frac{\Delta\phi_m}{\Delta t}$$

B = magnetic field
d = distance
ξ = emf
I = current
P = power
q = point charge
r = distance
U = potential (stored) energy
v = velocity or speed
Φ_m = magnetic flux
A = area
C = capacitance
E = electric field
F = force
ℓ = length
Q = charge
R = resistance
t = time
V = electric potential or potential difference
ρ = resistivity

KAPLAN 265

AP Physics

PRACTICE TEST I FORMULA SHEET (FREE RESPONSE SECTION ONLY)

GEOMETRY AND TRIGONOMETRY

Right Triangle

$$a^2 + b^2 = c^2$$

$$\sin \theta = \frac{a}{c}$$

$$\cos \theta = \frac{b}{c}$$

$$\tan \theta = \frac{a}{b}$$

Parallelepiped

$$V = \ell wh$$

Sphere

$$V = \frac{4}{3}\pi r^3$$

$$S = 4\pi r^2$$

Rectangle

$$A = bh$$

Circle

$$A = \pi r^2$$

$$C = 2\pi r$$

Cylinder

$$V = \pi r^2 \ell$$

$$S = 2\pi r \ell + 2\pi r^2$$

Triangle

$$A = \frac{1}{2}bh$$

C = circumference
S = surface area
h = height
w = width
A = area
V = volume
b = base
ℓ = length
r = radius

WAVES AND OPTICS

$$n = \frac{c}{v}$$

$$\sin \theta_c = \frac{n_2}{n_1}$$

$$M = \frac{h_i}{h_0} = -\frac{s_i}{s_0}$$

$$d \sin \theta = m\lambda$$

$$v = f\lambda$$

$$n_1 \sin \theta_1 = n_2 \sin \theta_2$$

$$\frac{1}{s_i} + \frac{1}{s_0} = \frac{1}{f}$$

$$f = \frac{R}{2}$$

$$x_m \approx \frac{m\lambda L}{d}$$

f = frequency or focal length
L = distance
m = an integer
R = radius of curvature
v = speed
λ = wavelength
d = separation
h = height
M = magnification
n = index of refraction
s = distance
x = position
θ = angle

Full-Length Practice Tests

PRACTICE TEST I FORMULA SHEET (FREE RESPONSE SECTION ONLY)

NEWTONIAN MECHANICS

$$x = x_0 + v_0 t + \frac{1}{2} a t^2$$

$$\sum \mathbf{F} = \mathbf{F}_{net} = m\mathbf{a}$$

$$a_c = \frac{v^2}{r}$$

$$\mathbf{p} = m\mathbf{v}$$

$$K = \frac{1}{2} m v^2$$

$$W = \mathbf{F} \cdot \Delta \mathbf{r} = F \Delta r \cos \theta$$

$$P = \mathbf{F} \cdot \mathbf{v} = Fv \cos \theta$$

$$U_s = \frac{1}{2} k x^2$$

$$T_p = 2\pi \sqrt{\frac{\ell}{g}}$$

$$F_G = \frac{G m_1 m_2}{r^2}$$

$$v = v_0 + at$$

$$v^2 = v_0{}^2 + 2a(x - x_0)$$

$$F_{fric} \le \mu N$$

$$\tau = rF \sin \theta$$

$$\mathbf{J} = \mathbf{F} \Delta t = \Delta \mathbf{p}$$

$$\Delta U_g = mgh$$

$$P_{avg} = \frac{W}{\Delta t}$$

$$\mathbf{F}_s = -k\mathbf{x}$$

$$T_s = 2\pi \sqrt{\frac{m}{k}}$$

$$T = \frac{1}{f}$$

$$U_G = -\frac{G m_1 m_2}{r}$$

F = force
h = height
K = kinetic energy
ℓ = length
N = normal force
p = momentum
$\mathbf{r}$ = position vector
t = time
v = velocity or speed
x = position
θ = angle
a = acceleration
f = frequency
J = impulse
k = spring constant
m = mass
P = power
r = radius or distance
T = period
U = potential energy
W = work done on a
$\quad$ system
μ = coefficient of
$\quad$ friction
τ = torque

KAPLAN 267

PRACTICE TEST I
ANSWER SHEET

1 Ⓐ Ⓑ Ⓒ Ⓓ Ⓔ
2 Ⓐ Ⓑ Ⓒ Ⓓ Ⓔ
3 Ⓐ Ⓑ Ⓒ Ⓓ Ⓔ
4 Ⓐ Ⓑ Ⓒ Ⓓ Ⓔ
5 Ⓐ Ⓑ Ⓒ Ⓓ Ⓔ
6 Ⓐ Ⓑ Ⓒ Ⓓ Ⓔ
7 Ⓐ Ⓑ Ⓒ Ⓓ Ⓔ
8 Ⓐ Ⓑ Ⓒ Ⓓ Ⓔ
9 Ⓐ Ⓑ Ⓒ Ⓓ Ⓔ
10 Ⓐ Ⓑ Ⓒ Ⓓ Ⓔ
11 Ⓐ Ⓑ Ⓒ Ⓓ Ⓔ
12 Ⓐ Ⓑ Ⓒ Ⓓ Ⓔ
13 Ⓐ Ⓑ Ⓒ Ⓓ Ⓔ
14 Ⓐ Ⓑ Ⓒ Ⓓ Ⓔ
15 Ⓐ Ⓑ Ⓒ Ⓓ Ⓔ
16 Ⓐ Ⓑ Ⓒ Ⓓ Ⓔ
17 Ⓐ Ⓑ Ⓒ Ⓓ Ⓔ
18 Ⓐ Ⓑ Ⓒ Ⓓ Ⓔ

19 Ⓐ Ⓑ Ⓒ Ⓓ Ⓔ
20 Ⓐ Ⓑ Ⓒ Ⓓ Ⓔ
21 Ⓐ Ⓑ Ⓒ Ⓓ Ⓔ
22 Ⓐ Ⓑ Ⓒ Ⓓ Ⓔ
23 Ⓐ Ⓑ Ⓒ Ⓓ Ⓔ
24 Ⓐ Ⓑ Ⓒ Ⓓ Ⓔ
25 Ⓐ Ⓑ Ⓒ Ⓓ Ⓔ
26 Ⓐ Ⓑ Ⓒ Ⓓ Ⓔ
27 Ⓐ Ⓑ Ⓒ Ⓓ Ⓔ
28 Ⓐ Ⓑ Ⓒ Ⓓ Ⓔ
29 Ⓐ Ⓑ Ⓒ Ⓓ Ⓔ
30 Ⓐ Ⓑ Ⓒ Ⓓ Ⓔ
31 Ⓐ Ⓑ Ⓒ Ⓓ Ⓔ
32 Ⓐ Ⓑ Ⓒ Ⓓ Ⓔ
33 Ⓐ Ⓑ Ⓒ Ⓓ Ⓔ
34 Ⓐ Ⓑ Ⓒ Ⓓ Ⓔ
35 Ⓐ Ⓑ Ⓒ Ⓓ Ⓔ
36 Ⓐ Ⓑ Ⓒ Ⓓ Ⓔ

37 Ⓐ Ⓑ Ⓒ Ⓓ Ⓔ
38 Ⓐ Ⓑ Ⓒ Ⓓ Ⓔ
39 Ⓐ Ⓑ Ⓒ Ⓓ Ⓔ
40 Ⓐ Ⓑ Ⓒ Ⓓ Ⓔ
41 Ⓐ Ⓑ Ⓒ Ⓓ Ⓔ
42 Ⓐ Ⓑ Ⓒ Ⓓ Ⓔ
43 Ⓐ Ⓑ Ⓒ Ⓓ Ⓔ
44 Ⓐ Ⓑ Ⓒ Ⓓ Ⓔ
45 Ⓐ Ⓑ Ⓒ Ⓓ Ⓔ
46 Ⓐ Ⓑ Ⓒ Ⓓ Ⓔ
47 Ⓐ Ⓑ Ⓒ Ⓓ Ⓔ
48 Ⓐ Ⓑ Ⓒ Ⓓ Ⓔ
49 Ⓐ Ⓑ Ⓒ Ⓓ Ⓔ
50 Ⓐ Ⓑ Ⓒ Ⓓ Ⓔ
51 Ⓐ Ⓑ Ⓒ Ⓓ Ⓔ
52 Ⓐ Ⓑ Ⓒ Ⓓ Ⓔ
53 Ⓐ Ⓑ Ⓒ Ⓓ Ⓔ
54 Ⓐ Ⓑ Ⓒ Ⓓ Ⓔ

55 Ⓐ Ⓑ Ⓒ Ⓓ Ⓔ
56 Ⓐ Ⓑ Ⓒ Ⓓ Ⓔ
57 Ⓐ Ⓑ Ⓒ Ⓓ Ⓔ
58 Ⓐ Ⓑ Ⓒ Ⓓ Ⓔ
59 Ⓐ Ⓑ Ⓒ Ⓓ Ⓔ
60 Ⓐ Ⓑ Ⓒ Ⓓ Ⓔ
61 Ⓐ Ⓑ Ⓒ Ⓓ Ⓔ
62 Ⓐ Ⓑ Ⓒ Ⓓ Ⓔ
63 Ⓐ Ⓑ Ⓒ Ⓓ Ⓔ
64 Ⓐ Ⓑ Ⓒ Ⓓ Ⓔ
65 Ⓐ Ⓑ Ⓒ Ⓓ Ⓔ
66 Ⓐ Ⓑ Ⓒ Ⓓ Ⓔ
67 Ⓐ Ⓑ Ⓒ Ⓓ Ⓔ
68 Ⓐ Ⓑ Ⓒ Ⓓ Ⓔ
69 Ⓐ Ⓑ Ⓒ Ⓓ Ⓔ
70 Ⓐ Ⓑ Ⓒ Ⓓ Ⓔ

Full-Length Practice Tests

PRACTICE TEST I

PRACTICE TEST I
AP PHYSICS B

Section I—Multiple Choice

Time: 90 minutes
Questions: 70
Points: 90

Directions: Each of the questions below is followed by five answer choices. Select the best answer, and fill in the corresponding oval on your answer sheet. This portion of the test should be taken without a calculator or reference to formulas.

Note: 10 m/s^2 may be used for g in calculations.

1. Which of the following statements is/are always true for completely inelastic collisions?

    I. System total mechanical energy is not conserved.

    II. System momentum is conserved.

    III. The objects stick together.

(A) I
(B) II
(C) I and II
(D) I and III
(E) I, II, and III

2. While rising vertically at 10 m/s, a helicopter passenger falls out the door. Fortunately, he has a rope tied to his harness so he can be retrieved. After 2 seconds, how much rope has been pulled out of the helicopter?

(A) 1 m
(B) 2 m
(C) 10 m
(D) 20 m
(E) 30 m

3. Which of the following is/are conserved in a totally inelastic collision?

(A) only momentum
(B) only kinetic energy
(C) both momentum and kinetic energy
(D) neither momentum nor kinetic energy
(E) It depends on the objects.

4. A spring is compressed a distance of 0.10 m from its rest position and held in place while a ball of mass 0.10 kg is placed at its end. When the spring is released, the ball leaves the spring traveling at 10 m/s. What is the spring constant?

(A) 10 N/m
(B) 50 N/m
(C) 100 N/m
(D) 500 N/m
(E) 1,000 N/m

GO ON TO THE NEXT PAGE. ➡

KAPLAN 269

AP Physics
PRACTICE TEST I

5. On a curved roadway of radius 100 m, the suggested speed limit is 25 mph (approximately 10 m/s). What must the coefficient of friction be between a car's tires and the road for the car to safely negotiate the curve at the posted speed?

   (A) 0.1
   (B) 0.2
   (C) 0.3
   (D) 0.4
   (E) 0.5

6. Which of the following arrangements of three masses (hanging off the horizontal rod) would result in a balanced mobile? Note: $M_1 = M_2 = M_3$.

   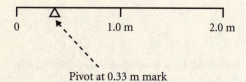

   Pivot at 0.33 m mark

	Mass 1	Mass 2	Mass 3
(A)	x = 0.5 m	x = 0.0 m	x = 1.5 m
(B)	x = 1.0 m	x = 0.0 m	x = 2.0 m
(C)	x = 2.0 m	x = 1.0 m	x = 0.0 m
(D)	x = 0.0 m	x = 1.0 m	x = 0.0 m
(E)	x = 0.5 m	x = 0.0 m	x = 2.0 m

7. In the diagram below, a picture frame held by two ropes is at rest. Find the value of the tension in rope no. 1.

   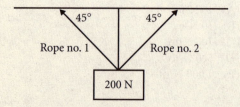

   (A) 50 N
   (B) 70 N
   (C) 100 N
   (D) 140 N
   (E) 200 N

8. Two skaters stand face to face. Skater 1 has a mass of 55 kg, and skater 2 has a mass of 75 kg. They push off one another and move in opposite directions. What is the ratio of skater 1's speed to skater 2's speed?

   (A) 1 : 1
   (B) 1 : 2
   (C) 55 : 75
   (D) 55 : 130
   (E) 75 : 55

9. The drawing shows the path of a projectile that was launched at an angle from the ground. At which points would you find an acceleration of zero, a maximum speed, and maximum height?

   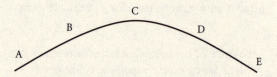

	a = 0	v = max	height = max
(A)	B	E	E
(B)	no point	C	D
(C)	C	B	C
(D)	no point	A	C
(E)	E	A	C

10. An amusement park ride has a diameter of 16 meters and revolves at 15 rpm (revolutions per minute). How fast is a rider moving if she sits at the outermost part of the ride?

    (A) 29.4 m/s
    (B) 1.07 m/s
    (C) 225 m/s
    (D) 12.56 m/s
    (E) 9.8 m/s

GO ON TO THE NEXT PAGE.

11. How much total mechanical work does a weight lifter do in 10 repetitions of lifting (and lowering) 200 kg to a height of 1.0 m?

    (A) zero
    (B) 200 J
    (C) 2,000 J
    (D) 19,600 J
    (E) 39,200 J

12. A 500 kg dragster finishes a race and, finding that its brakes have failed completely, engages a parachute from the rear of the vehicle to slow down. Initially, it is traveling at 70 m/s, and in 4 seconds it has slowed down to 30 m/s. What is the average force exerted on the car by the parachute?

    (A) 150 N
    (B) 300 N
    (C) 500 N
    (D) 1,000 N
    (E) 5,000 N

13. A massless string connects three blocks as shown below. A force of 12 N acts on the system. What is the acceleration of the blocks and the tension in the rope attached to the 1 kg block?

    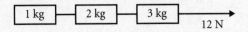

	Acceleration	Tension
(A)	4 m/s^2	6 N
(B)	2 m/s^2	2 N
(C)	2 m/s^2	12 N
(D)	12 m/s^2	6 N
(E)	4 m/s^2	3 N

14. The diagram shows three positions of a mass as it bounces on a spring. The spring is attached to a ceiling. What is true at point C, the lowest point of the oscillation?

    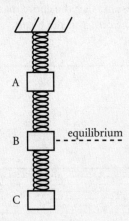

    I. Speed = 0 m/s
    II. Acceleration = 0 m/s^2
    III. Displacement = 0 m

    (A) I
    (B) II
    (C) III
    (D) I and II
    (E) II and III

15. What is the gravitational force between two runners standing 100 meters apart on a track if their masses are 60 kg and 75 kg respectively?

    (A) 588 N for the first runner and 735 N for the second runner
    (B) 45 N
    (C) $3 \cdot 10^{-7}$ N
    (D) $3 \cdot 10^{-11}$ N
    (E) 3.0 N

GO ON TO THE NEXT PAGE.

16. In the diagram below, how far from the end of the board can a 60 kg person walk before it will tip over? The mass of the board is 100 kg, and a 2-meter section of its 6-meter length hangs over the edge. Assume that the entire mass of the board acts at the center of the board.

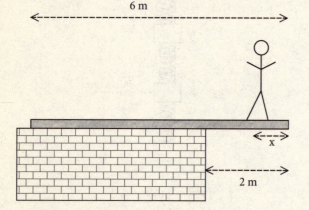

(A) 1.33 meters
(B) 1 meter
(C) 2 meters
(D) 0.33 meters
(E) 0.5 meters

17. Which force in the diagram will produce the largest torque about the pivot shown?

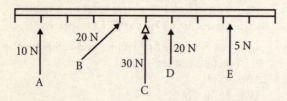

(A) A
(B) B
(C) C
(D) D
(E) E

18. How much energy is stored in a spring (k = 300 N/m) when it is compressed by 0.05 m from its rest position?

(A) 6.0 J
(B) 1.5 J
(C) 0.375 J
(D) 60,000 J
(E) 15 J

19. In the picture below, where should a third mass (6 kg) be hung to produce rotational equilibrium?

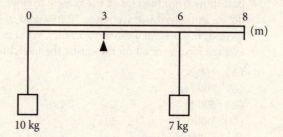

(A) 2 meters to the right of the fulcrum
(B) 1.88 meters to the left of the fulcrum
(C) 1.5 meters to the right of the fulcrum
(D) 0.17 meters to the left of the fulcrum
(E) There is no place to put the mass so that it will balance.

20. All of the following are true about torque EXCEPT:

(A) Its units are Newton-meters.
(B) It can only be produced by a force.
(C) Torque is zero unless the force has a component perpendicular to the moment arm.
(D) A force whose line of action passes through the axis of rotation produces zero torque.
(E) Torque may be clockwise or counterclockwise, depending on the direction the force points.

GO ON TO THE NEXT PAGE.

21. How much work is done by the centripetal force (in this case, the friction between the tires and the road) on a 1,200 kg car moving on a circular track of radius 50 meters at a constant speed of 31.4 m/s?

    (A) 7.4×10^6 J
    (B) 19.7 J
    (C) 2,136.7 J
    (D) 117,600 J
    (E) 0 J

22. In which of the following collisions is total mechanical energy NOT conserved?

    I. elastic
    II. partially inelastic
    III. completely inelastic

    (A) I only
    (B) III only
    (C) I and II only
    (D) II and III only
    (E) I, II, and III

23. A runner runs the first half of a 10 km race at a steady speed of 5 m/s. Due to fatigue, she slows her pace to a constant 4 m/s for the second half. What is her average speed for the race?

    (A) 3.5 m/s
    (B) 4.4 m/s
    (C) 4.5 m/s
    (D) 5.0 m/s
    (E) 5.5 m/s

24. In the apparatus shown below, what is the tension in the rope, if we assume constant velocity?

    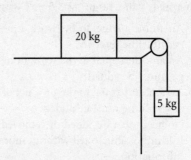

    (A) 39 N
    (B) 1.9 N
    (C) 50 N
    (D) 78 N
    (E) 100 N

25. A piece of solid lead weighing 6.5 kg at 327°C (its melting point) receives 150,000 joules of heat. In what state is the lead after this occurs? The latent heat of fusion for lead is 23,200 J/kg.

    (A) 6.5 kg of solid lead at 327°C
    (B) 6.5 kg of solid lead at a higher temperature
    (C) 6.5 kg of melted lead at 327°C
    (D) Some of it is melted lead, and some of it stays solid.
    (E) 6.5 kg of melted lead at a higher temperature

26. Which property describes the change in pitch that you hear as you move closer to a sound?

    (A) resonance
    (B) decibels
    (C) open/closed pipes
    (D) Doppler effect
    (E) sonic boom

AP Physics
PRACTICE TEST I

27. How much heat must be added or removed to allow 2 moles of an ideal gas to expand isothermally from 1.0 m^3 to 3.0 m^3?

    (A) An amount equal to the work done by the gas must be removed.
    (B) An amount equal to the work done by the gas must be added.
    (C) An amount equal to the product of pressure and volume must be added.
    (D) No heat must be added or removed: Q = 0.
    (E) It is impossible to tell without more information.

28. An open tube and a closed tube resonate at a fundamental frequency f. What is the next higher frequency at which each tube will resonate?

	Open tube	Closed tube
(A)	f	$2f$
(B)	$2f$	$2f$
(C)	$3f$	$3f$
(D)	$2f$	$4f$
(E)	$2f$	$3f$

29. An engine receives 475 J from a heat reservoir at 750 K and rejects 315 J to the cold reservoir at 250 K. Find the actual efficiency and the theoretically maximal possible Carnot efficiency for this engine.

	Actual	Carnot
(A)	33.7%	66.7%
(B)	16.5%	50.0%
(C)	66.7%	66.7%
(D)	107.8%	75.0%
(E)	50.8%	75.0%

30. An ideal gas is held in a rigid container whose volume is 0.25 m^3. It is heated from 100°C, where it is at a pressure of 2×10^5 Pa, to a final temperature of 250°C. What is the new pressure of the gas?

    (A) 10×10^5 Pa
    (B) 2.8×10^5 Pa
    (C) 5.0×10^5 Pa
    (D) 0.80×10^5 Pa
    (E) 150×10^5 Pa

31. A mass is moving in uniform circular motion under the effects of a centripetal force. If the force is doubled and the radius is cut in half, by what factor will the speed change?

    (A) It will double.
    (B) It will be cut in half.
    (C) It will not change.
    (D) It will quadruple.
    (E) It will be $\sqrt{2}$ times faster.

32. How many electrons are transferred in the process of charging a latex balloon to 1.6×10^{-8} C?

    (A) 1×10^{11}
    (B) 2.56×10^{-27}
    (C) 1×10^{27}
    (D) 1×10^{-25}
    (E) Latex balloons can't be electrically charged.

33. A circular, rotating space station uses centripetal force to produce artificial gravity. If the radius of the space station is 450 m and it spins at a speed of 55 m/s, what is the apparent value of g on the surface of the station?

    (A) 1.8 m/s^2
    (B) 9.8 m/s^2
    (C) 13.1 m/s^2
    (D) 6.7 m/s^2
    (E) zero, because in space everything is weightless

GO ON TO THE NEXT PAGE.

34. On the spring shown below, at what respective value(s) of x, representing displacements from equilibrium, are the velocity and the acceleration 0?

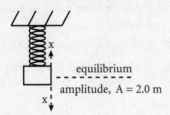

	Velocity	Acceleration
(A)	−2.0 m	2.0 m
(B)	0 m	2.0 m
(C)	0 m	0 m
(D)	2.0 m	0 m
(E)	2.0 m	-2.0 m

35. You are playing middle A (440 Hz) on the saxophone on a street corner. A trombone is playing middle A (440 Hz) in the open bed of a pickup truck as it approaches you. You hear a beat frequency of 4 Hz. What frequency do you hear from the trombone and what property of sound waves is being demonstrated in this problem?

	Frequency	Property
(A)	444 Hz	Doppler effect
(B)	444 Hz	Fundamental frequency
(C)	436 Hz	Doppler effect
(D)	436 Hz	Harmonics
(E)	436 Hz	Speed of sound in air

36. A 343 Hz note is played outdoors on a day when the temperature is 25°C. What is the wavelength of this note?

(A) 0.985 m
(B) 1.01 m
(C) 13.3 m
(D) 215 m
(E) 343 m

37. On an unknown planet, a pendulum of length 1.5 m swings with a period of 3.3 s. What is the acceleration due to gravity on this planet?

(A) 3.7 m/s^2
(B) 12.7 m/s^2
(C) 7.1 m/s^2
(D) 5.4 m/s^2
(E) 9.8 m/s^2

38. The fundamental frequency of a closed pipe is 108 Hz. Find the second overtone for this tube.

(A) 540 Hz
(B) 324 Hz
(C) 432 Hz
(D) 27 Hz
(E) 1,080 Hz

39. A 1.0 m long pendulum on a distant planet swings in simple harmonic motion with a period of 3.0 s. What is the approximate magnitude of the acceleration of gravity at this location?

(A) 0 m/s^2
(B) 2.0 m/s^2
(C) 4.0 m/s^2
(D) 8.0 m/s^2
(E) 16 m/s^2

40. What is the height of the image formed by placing a 6 cm tall object a distance 20 cm in front of a concave mirror with a focal length of 10 cm?

(A) 3 cm
(B) 6 cm
(C) 8 cm
(D) 10 cm
(E) 12 cm

AP Physics
PRACTICE TEST I

41. If the mass on an oscillating spring is decreased to $\frac{1}{2}$ the original mass:

    (A) The period will be cut in half, but the frequency will remain the same.
    (B) The period and the frequency will remain the same.
    (C) The period will be increased by the square root of two.
    (D) The period will be decreased by a factor of two.
    (E) The frequency will be increased by a factor of the square root of two.

42. What is the best definition of capacitance?

    (A) capacity to do work
    (B) how much charge a capacitor can hold
    (C) how much charge can be stored per volt of potential difference
    (D) how large a capacitor is
    (E) none of the above

43. What is the potential energy of a 3×10^{-6} C charge when it's placed at a distance of 2.5 cm from another 3×10^{-6} C charge?

    (A) −202.5 J
    (B) 0.032 J
    (C) 130 J
    (D) 0.096 J
    (E) 3.2 J

44. What is the net force on the particle on q_3?

    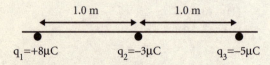

    (A) 90 mN to the left
    (B) 90 mN to the right
    (C) 45 mN to the left
    (D) 45 mN to the right
    (E) 2.25 mN to the right

45. Which of the following is/are characteristic of both Coulomb's law and the universal law of gravitation?

    I. It describes attractive and repulsive forces.
    II. It obeys an inverse square law.
    III. It describes field forces (they don't require contact between objects).

    (A) I only
    (B) III only
    (C) I and II
    (D) II and III
    (E) I, II, and III

46. Where in the diagram below is a region in which there is at least one point where the electric field is equal to zero?

    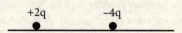

    (A) to the left of +2q
    (B) between +2q and −4q
    (C) to the right of −4q
    (D) above and below the line on which +2q and −4q lie
    (E) there is no point where E = 0.

47. What is the potential difference between points A and B?

    E = 100N/C, AB = 0.2 m

    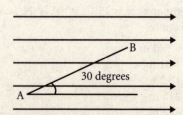

    (A) 250 V
    (B) 10.0 V
    (C) 20.0 V
    (D) 17.3 V
    (E) not enough information to answer

GO ON TO THE NEXT PAGE.

48. How much work is done in transferring 3 C of charge between the terminals of a AA (1.5 V) battery?

   (A) 1.5 J
   (B) 2.0 J
   (C) 3.375 J
   (D) 4.5 J
   (E) 0.5 J

49. An electron enters a magnetic field from the left, as shown below. Which of the following best describes the path of the electron as it travels through the field?

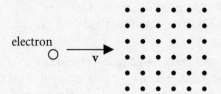

   (A) It travels straight through the field undisturbed.
   (B) It travels in a line that moves downward at an angle.
   (C) It travels in a circle in a clockwise direction.
   (D) It travels in the direction the magnetic field points.
   (E) It travels in a circle in a counter-clockwise direction.

50. Which of the following is the best example of Lenz's law?

   (A) A magnet dropped down a copper tube will accelerate at a rate greater than gravity.
   (B) A heating element doesn't heat as fast as it should due to the magnetic forces on it.
   (C) Eyeglasses make a person see better.
   (D) A metal bar moving across a magnetic field will slow down unless a force is maintained.
   (E) An electron moving through a magnetic field will tend to move in a circular path.

51. Which compass pole points to a region near the geographic North Pole of the Earth?

   (A) The north compass pole
   (B) The south compass pole
   (C) It depends on the location of the compass on Earth.
   (D) A compass actually points toward the geographic North Pole of Earth.
   (E) Not enough information is given to determine the answer.

52. As shown below, a beam of protons is fired into a magnetic field (B = 1.5 T) at a velocity v = 2.0 · 10^6 m/s. Find the magnitude and direction of the force on each proton.

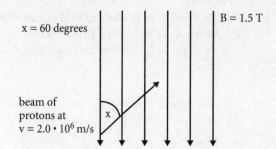

   (A) 4.2×10^{-13} N; out of the page
   (B) 4.2×10^{-13} N; into the page
   (C) 3.0×10^6 N; into the page
   (D) 4.8×10^{-10} N; into the page
   (E) 4.8×10^{-10} N; out of the page

53. All of the following are accurate expressions for the units of a magnetic field EXCEPT:

   (A) $\dfrac{N}{A \cdot m}$
   (B) $\dfrac{N}{C \cdot m/s}$
   (C) T
   (D) $\dfrac{V}{m^2/s}$
   (E) $\dfrac{kg \cdot s}{C \cdot m}$

54. As shown below, two parallel wires carry currents I_1 and I_2. Which statement is true about the wires?

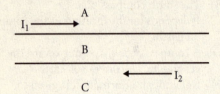

(A) They will not move unless an outside force acts on them.
(B) They will tend to move away from one another.
(C) One of the wires will tend to lift off the ground.
(D) They will tend to move toward one another.
(E) Whatever effect occurs, switching the direction of I_1 will have no impact on that effect.

55. Using the same diagram as question 54, where will the magnetic field NEVER be zero?

(A) Nowhere
(B) Section A
(C) Section B
(D) Section C
(E) You need to know the currents to answer this.

56. Let's say you have two long pieces of metal that, by sight, appear to be identical. However, one is a magnet, and the other is simply a piece of metal. How could you determine which is the magnet?

(A) Arrange the pieces in a T-shape. If they attract, the top of the T is the magnet.
(B) Arrange the pieces in a T-shape. If they attract, the top of the T is the metal.
(C) Arrange the pieces in a T-shape. If they don't attract, the top of the T is the metal.
(D) You cannot tell which is the magnet.
(E) You need more materials to answer this.

57. Describe the motion of the charged particle as it enters the magnetic field shown below:

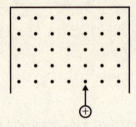

(A) curves into the page
(B) straight
(C) semicircular, to the left
(D) semicircular, to the right
(E) curves out of the page

58. A charge (q = +4.2 µC) is placed at x = −0.5 m. A second charge (q = −2 µC) is placed at the origin. Where along the x-axis is the electric field equal to zero?

(A) at the positive charge
(B) to the left of the positive charge
(C) to the right of the negative charge
(D) at the negative charge
(E) between the two charges

59. A circuit is wired with two 20-ohm resistors in parallel. When it is supplied with 10 volts:

(A) a current of 0.5 amp will flow in each resistor
(B) a 5-volt drop will occur across each resistor
(C) a current of 1 amp will flow in each resistor
(D) a 20-volt drop will occur across each resistor
(E) a current of 0.25 amp will flow in each resistor

GO ON TO THE NEXT PAGE.

Full-Length Practice Tests

PRACTICE TEST I

60. When comparing the fission of a uranium nucleus to the fusion of two deuterium nuclei, which statement is true about the amount of energy released?

(A) Fission of a uranium nucleus releases the most energy.

(B) Fusion of two deuterium nuclei releases the most energy.

(C) They release equal amounts of energy.

(D) The energy released depends on the ambient temperature at the time of the reaction.

(E) The energy released depends on how much material is involved in the reaction.

61. Which decay produces a positron and drops a nucleus to the next lower atomic number?

(A) alpha

(B) beta$^-$

(C) beta$^+$

(D) gamma

(E) fission

62. Why is a sustained fusion reaction impossible to achieve at room temperature?

(A) The nuclei only move at high temperatures.

(B) The nuclei won't break into smaller nuclei at room temperature.

(C) Hydrogen can't be contained at room temperature.

(D) Nuclei move too slowly at room temperature.

(E) Nuclei won't fuse together at any temperature.

63. What nuclear transformation occurs during beta$^-$ decay?

(A) A proton decays into a neutron and an electron.

(B) An electron and proton combine to form a neutron.

(C) A neutron decays into a proton and an electron plus a neutrino.

(D) An electron and a neutron combine to form a proton.

(E) A proton and a neutron combine to form a positron.

64. How much energy is produced when $_{92}^{235}$U is bombarded with a neutron and undergoes a fission reaction to produce $_{54}^{140}$Xe, $_{38}^{94}$Sr, and two neutrons, creating a mass defect of 0.198 atomic mass units?

(A) 0.198 MeV

(B) 2.21 MeV

(C) 931.5 MeV

(D) 46.4 MeV

(E) 184.7 MeV

65. 224-Radium has a half-life of 3.66 days. What can you say about one nucleus of radium after 3.66 days?

(A) Half of the nucleus will decay.

(B) It will only decay when a second nucleus is brought into contact with it.

(C) There is a 50 percent chance it will have decayed.

(D) It won't decay prior to 3.66 days, but should decay after that time.

(E) It will undergo alpha decay, but only produce $\frac{1}{2}$ of an alpha particle.

66. All of the following are conserved in a nuclear reaction EXCEPT:

(A) momentum

(B) energy

(C) nucleon number

(D) charge

(E) velocity

GO ON TO THE NEXT PAGE. ➡

KAPLAN 279

67. Two loops have current flowing through them as shown. The top loop lies directly above the bottom one. When the current in the bottom loop is increased, what happens to the current in the top loop?

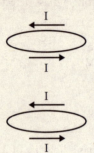

(A) It increases.
(B) It decreases.
(C) It drops to zero.
(D) It doesn't change.
(E) It depends on the relative values for I.

68. Which of the following units are used to measure electric forces of attraction between charges?

(A) Coulombs
(B) Amperes
(C) Newtons
(D) Ohms
(E) Farads

69. Which of the following nuclear reactions is impossible?

(A) $^{238}_{92}U \rightarrow \,^{234}_{90}Th + \,^{4}_{2}He$

(B) $^{212}_{83}Bi \rightarrow \,^{212}_{84}Po + \,^{0}_{1}e + \nu$

(C) $n + \,^{235}_{92}U \rightarrow \,^{140}_{54}Xe + \,^{94}_{38}Sr + 2n$

(D) $^{212}_{84}Po \rightarrow \,^{208}_{82}Pb + \alpha$

(E) $^{212}_{82}Pb \rightarrow \,^{211}_{83}Pb + \,^{0}_{1}e + \nu$

70. In a photoelectric effect experiment, which of the following changes may result in a current, if there was no current flowing previously?

(A) making the color of the incident light "bluer"
(B) increasing the wavelength of the incident light
(C) making the incident light brighter
(D) decreasing the frequency of the incident light
(E) decreasing the voltage in the apparatus

END OF SECTION I

PRACTICE TEST I
AP PHYSICS B

Section II—Free Response

Time—90 minutes
Questions—6
Points—90

Directions: This portion of the test consists of six free-response questions and should be taken in a 90-minute time period. The point value of each question is indicated. Not all parts of each question have equal weight. A calculator and list of formulas are allowed.

1. (15 points)

    Consider the pulley apparatus below. A box of mass 10 kg is being pulled across a table by a falling box of mass 7.5 kg that is suspended by a cord over a frictionless, massless pulley. The coefficient of kinetic friction between the box and table is 0.20.

    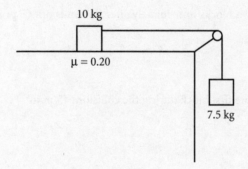

    a) Find the acceleration of the two-block system. (3 points)

    b) What is the tension in the cord? (3 points)

    Now suppose the hanging mass is <u>replaced</u> by a downward force of 75 N.

    c) What will be the resulting acceleration of the system? (3 points)

AP Physics
PRACTICE TEST I

d) What is the new tension in the cord? (3 points)

e) How much mass should be added to the first block for it to slide at constant velocity? (3 points)

2. (15 points)

A block slides along a frictionless surface toward a second block attached to the end of a spring with elastic constant 250 N/m, as shown below. The first block slides at a speed of 8 m/s and the second block is initially stationary. When the blocks collide they stick together.

$$V = 8 \text{ m/s} \qquad v = 0 \text{ m/s} \qquad k = 250 \text{ N/m}$$
$$m = 5 \text{ kg} \qquad m = 2 \text{ kg}$$

a) What is the speed of the two blocks immediately after the collision? (5 points)

b) What impulse is given to the first block during the collision? (5 points)

c) How far will the spring compress? (3 points)

d) What is the maximum speed the blocks will attain once they begin oscillating on the spring? (2 points)

GO ON TO THE NEXT PAGE.

282   **KAPLAN**

Full-Length Practice Tests

PRACTICE TEST I

3. (15 points)

A 50 g piece of ice at an initial temperature that is unknown is placed into a glass containing 400 g of water at 20°C. The glass and ice come to thermal equilibrium at 5°C.

a) The piece of ice initially floats with 90 percent of its volume under the water surface.

    (i) What is the specific gravity of the ice? Justify your answer. (2 points)

    (ii) What is the buoyant force on the ice? (3 points)

b) As the ice melts, assume that no heat is lost to the container or surroundings. Find the initial temperature of the ice. Assume the specific heat of ice is 2,000 J/kg•°C, the specific heat of water is 4,180 J/kg•°C, and the latent heat of fusion of ice is $3.35 \cdot 10^5$ J/kg. (5 points)

c) Now suppose that a large block of the original ice (mass = 5 kg) has a 0.75 kg piece of hot (200°C) metal placed on it so that in time, 0.27 kg of the ice melts. What is the specific heat capacity of the unknown metal? (5 points)

4. (15 points)

A musician blows into an open pipe and produces its fundamental frequency. The pipe is 0.70 meters long. (Use $v_{sound} = 343$ m/s.)

a) What is the fundamental frequency? (5 points)

b) If he wanted to play a higher note, what would be the next higher frequency that can be played on this pipe? (4 points)

c) How many wavelengths fit inside the tube in this case? (2 points)

d) If the musician were to replace his open pipe with a closed pipe, what is one possible length for the pipe if he wants to produce the same (higher) frequency? (4 points)

GO ON TO THE NEXT PAGE. ➡

KAPLAN   283

AP Physics
PRACTICE TEST I

5. (15 points)

An object of height 5 cm is located 8 cm from a converging lens of focal length 12 cm, as shown below.

a) On the diagram below, locate the image formed by drawing at least two principal rays. (6 points)

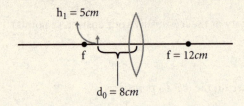

b) Calculate the image distance. (3 points)

c) What kind of image is this? Clearly explain how you know this. (2 points)

d) Calculate the magnification of this lens. (2 points)

e) Calculate the height of the image formed. (2 points)

GO ON TO THE NEXT PAGE.

6. (15 points)

In the diagram below, an elementary charge is accelerated between two charged plates into a magnetic field. The mass of a proton = 1.67×10^{-27} kg. The mass of an electron = 9.11×10^{-31} kg.

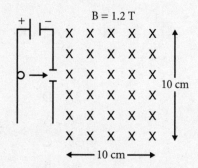

a) What is the sign of the elementary charge? (2 points)

b) If the charge is traveling at a speed of 2.0×10^5 m/s when it passes through the opening, what is the voltage charging the two plates? (4 points)

c) Describe the path of the charge in the magnetic field. (4 points)

d) Find the distance from the entrance point to where the charge leaves the magnetic field. (5 points)

END OF SECTION II

PRACTICE TEST I
ANSWER KEY

Section I—Multiple-Choice

1. E	19. C	37. D	55. C
2. D	20. B	38. A	56. B
3. A	21. E	39. C	57. D
4. E	22. D	40. B	58. C
5. A	23. B	41. E	59. A
6. D	24. C	42. C	60. A
7. D	25. D	43. E	61. C
8. E	26. D	44. D	62. D
9. D	27. B	45. D	63. C
10. D	28. E	46. A	64. E
11. A	29. A	47. D	65. C
12. E	30. B	48. D	66. E
13. B	31. C	49. E	67. B
14. A	32. A	50. D	68. C
15. D	33. D	51. A	69. E
16. D	34. D	52. B	70. A
17. A	35. A	53. E	
18. C	36. B	54. B	

Answers and Explanations to Practice Test I

SECTION I: MULTIPLE-CHOICE

Note: 10 m/s^2 is used for g in most cases.

1. **(E)** For an inelastic collision, momentum is still conserved, but the mechanical energy is converted to thermal energy during the impact. Assuming inelastic implies totally inelastic, the objects also stick together.

2. **(D)** The helicopter's displacement upward:

    $$s = v_0 t + \frac{1}{2} at^2 = (10 \text{ m/s})(2 \text{ s}) = 20 \text{ m upward}$$

    The passenger's displacement:

    $$s = (10 \text{ m/s})(2 \text{ s}) + \frac{1}{2}(-10 \text{m/s}^2)(2 \text{ s})^2 = 0$$

    Therefore, after two seconds a distance of 20 meters separates them, and thus 20 m of rope has been used.

3. **(A)** Only momentum is conserved in totally inelastic collisions.

4. **(E)** The loss of potential energy of the spring is equal to the gain in kinetic energy of the ball, noting that the ball is released from the spring at equilibrium. As the ball passes equilibrium, the ball continues moving and the spring begins to compress again.

    $$\frac{1}{2} kx^2 = \frac{1}{2} mv^2$$
    $$k = mv^2/x^2 = (0.10 \text{ kg})(10 \text{ m/s})^2/(0.10)^2 = 1{,}000 \text{ N/m}.$$

KAPLAN 287

AP Physics
ANSWERS AND EXPLANATIONS TO PRACTICE TEST I

5. **(A)** The centripetal force to keep the car in a level turn is provided by the friction force, which equals coefficient of friction times the normal force. On a level turn, the normal force is equal to the weight of the car, so:

$$F_c = F_f = \mu N = \mu mg$$
$$mv^2/R = \mu mg$$
$$\mu = v^2/gR = (10 \text{ m/s})^2/ (10 \text{ m/s}^2)(100 \text{ m}) = 0.1.$$

6. **(D)** Since the three masses are equal and the pivot is at the 0.33 meter position, the solution must include distances that produce balanced torque on each side of the pivot.

$$\tau_{clockwise} = \tau_{counterclockwise}, \text{ where } \tau = (\text{force})(\text{lever arm})$$

Answer **(D)** produces: $M(0.33 \text{ m}) + M(0.33 \text{ m}) = M(0.67 \text{ m})$.

7. **(D)** Due to symmetry, each rope is supporting the same amount. Thus, the vertical component of each rope is supporting half of the frame, or 100N. Using right triangle trigonometry,

$$T \sin45° = 100 \text{ N}$$
$$T = 100/\sin45° = \frac{100(2)}{\sqrt{2}}$$

The most reasonable answer is 140 N.

8. **(E)** Since momentum must be conserved and the total initial momentum is zero, the skaters' final momenta must be equal in magnitude and opposite in direction in order to add to zero afterward. Using only absolute values:

$$m_1V_1 = m_2V_2 \quad \text{and} \quad V_1/V_2 = m_2/m_1 = 75 : 55$$

9. **(D)** The projectile has a constant force of gravity on it, so acceleration is never 0. Maximum velocity occurs near the ground, at points A and E, and maximum height is obviously at C.

288 **KAPLAN**

AP Physics

ANSWERS AND EXPLANATIONS TO PRACTICE TEST I

10. **(D)** To calculate the linear velocity at the outside, use $v = \omega R$, where R is 8 meters and ω is 15 revolutions per minute. The angular velocity, ω, must first be converted to radians per second:

$$(15 \text{ rev/min})(2\pi \text{ rad/rev})/(60 \text{ s/min}) = \pi/2 \text{ rad/s}$$

Then solve for the linear velocity:

$$v = (\pi/2 \text{ rad/s})(8 \text{ m}) = 4\pi \text{ m/s}$$

This answer is closest to choice **(D)**, 12.56 m/s.

11. **(A)** If the weight lifter lifts and lowers the mass, regardless of how many times, the mass returns to its starting position. The resultant displacement against gravity is zero, so no work is done ($W = F \bullet s$).

12. **(E)** Using Newton's Second Law and substituting the formula for acceleration:

$$F = ma = m(v_f - v_o/t) = (500 \text{ kg})[(30 \text{ m/s} - 70 \text{ m/s})(4 \text{ s})] = -5000 \text{ N}$$

or 5,000 Newtons opposing the motion.

13. **(B)** Using Newton's Second Law on the system of blocks:

$$\Sigma F = m_{total}a$$
$$12 \text{ N} = (6 \text{ kg})(a)$$
$$a = 2 \text{ m/s}^2$$

The acceleration is the same for all the blocks.

Applying the law to only the 1 kg block and using the calculated acceleration:

$$T = (1\text{kg})(a) = (1 \text{ kg})(2 \text{ m/s}^2)$$
$$T = 2 \text{ N}.$$

14. **(A)** At the lowest point of its oscillation, the spring has converted all its energy to potential energy, so speed is zero. At this point also, there is maximum displacement. Since the spring force is greatest at maximum displacement, the acceleration is also maximum, by Newton's second law.

KAPLAN 289

AP Physics
ANSWERS AND EXPLANATIONS TO PRACTICE TEST I

15. **(D)** $F_G = G\, m_1 m_2/R^2 = (6.67 \times 10^{-11})(60\text{ kg})(75\text{ kg})/(100\text{ m})^2$

    Since the answer choices are widely separated in value, a quick estimate by cancelling the 60 and 75 with the 100's lets you see that the answer is nearest to **(D)**.

16. **(D)** Placing the pivot at the edge of the building and balancing clockwise and counterclockwise torques:

    (weight of board)(distance of center of board from pivot) =

    (weight of walker)(distance from pivot)

    $(100\text{ kg})(g)(1\text{ m}) = (60\text{ kg})(g)(2 - x)$, where x is distance of walker from far end of board

    $x = 0.33$ m

17. **(A)** Remember that only the force component perpendicular to the lever arm distance will produce a torque. Also remember that a force applied at the pivot will produce no torque. Thus, the force at **(A)** has the largest product of force and lever arm.

18. **(C)** Potential energy of a spring $= \dfrac{1}{2}\, kx^2 = \left(\dfrac{1}{2}\right)(300\text{ N/m})(0.05)^2$.

    To solve quickly without a calculator, write 0.05 as $\dfrac{5}{100}$ or $\dfrac{1}{20}$:

    $\left(\dfrac{1}{2}\right)(300)/(20)^2 = 300/800 = \dfrac{1}{8}\text{ J} = 0.375\text{ J}.$

19. **(C)** Torque is calculated by taking mass times g to get force and then multiplying by lever arm, which should be measured perpendicular from line of force to pivot. The 10 kg mass is 3 m from the pivot on the left, producing a 300 N·m torque counterclockwise. The 7 kg mass is also 3 m from the pivot, but on the right, producing a 210 N·m torque clockwise. Thus, an added 90 N·m clockwise torque is needed for equilibrium. A 6 kg mass placed 1.5 m to the right of the fulcrum would accomplish this.

20. **(B)** Torque can be produced by a force or by another torque.

21. **(E)** Since the centripetal force is always directed radially inward and the car's velocity is tangential, the force here is perpendicular to the direction of motion, so no work is done: $W = Fs\cos\theta$ or $W = F\bullet s$.

22. **(D)** Total mechanical energy is not conserved in partially or totally inelastic collisions.

290 **KAPLAN**

AP Physics

ANSWERS AND EXPLANATIONS TO PRACTICE TEST I

23. **(B)** Average speed is total distance divided by total time. The total distance run is 10,000 meters. The total time is the time for the first leg, 5,000 meters divided by 5 m/s, plus the time for the second leg, which is 5,000 meters divided by 4 m/s. The solution reduces to:

$$V = \frac{10,000 \text{ m}}{1,000_S + 1,250_S} = \frac{10,000 \text{ m}}{2,250_S}$$

24. **(C)** If we assume the system is moving at constant velocity, then it is in equilibrium, so the tension in the rope is equal to the hanging weight, which is 5 kg times g.

25. **(D)** The latent heat of fusion value tells us that 23,200 Joules would be required to melt just one kilogram of the lead. To completely melt 6.5 kilograms would require 6.5 times 23,200 Joules, and that amount of energy is not supplied. Therefore, only part of the lead will melt.

26. **(D)** As you move toward a sound source, the pitch or frequency of the source appears to rise. This is called the Doppler effect.

27. **(B)** No actual calculation is needed here. The term isothermal means there is no change in temperature, and thus no change in internal energy, U. Therefore, by the First Law of Thermodynamics, which is $Q = U - W$, if $U = 0$, then $Q = -W$. Since W is the work done by the gas and Q is heat added to the gas, we know that heat added will do negative work, or allow the gas to expand.

28. **(E)** For an open tube, since an antinode occurs at each end of the tube, the harmonics have wavelengths, $\lambda_n = 2L/n$. Thus, the second harmonic, $n = 2$, will have half the wavelength of the first. Since we assume the speed of each wave remains constant, by $v = f\lambda$, the frequency must double when the wavelength is halved.

For a closed tube, with a node at the closed end, the harmonics have wavelengths $\lambda_n = 4L/(2n-1)$. The wavelength for the second harmonic is $\frac{1}{3}$ the wavelength for the first, so the frequency of the second harmonic must be three times that of the first.

29. **(A)** Carnot efficiency is the difference between high and low temperature reservoirs divided by the high temperature, or $e_c = (750 - 250)/750 = 0.67$. Actual efficiency is work output divided by heat input, or $e = \frac{Q_H - Q_L}{Q_H} = \frac{160}{475} = 0.33,$ which leads us (without actually calculating the value) to answer **(A)**.

KAPLAN 291

AP Physics
ANSWERS AND EXPLANATIONS TO PRACTICE TEST I

30. **(B)** Since volume is held constant, pressure increases directly with temperature:

$$P_1/T_1 = P_2/T_2$$

Temperatures <u>must</u> be changed to Kelvin.

$$(2 \times 10^5 \text{ Pa})/373 \text{ K} = (P)/523 \text{ K}$$

Without actually working it out, we can see that the Kelvin temperature changes by a factor of about one and a half, so the new pressure must be between one and two times the original pressure. Thus, the answer we select is 2.8×10^5 pa.

31. **(C)** Since $F_c = mv^2/R$, doubling the force and cutting the radius in half would have an equivalent mathematical effect on the two sides of the equation. Thus, the speed will not change.

32. **(A)** Each electron has a charge of 1.6×10^{-19} Coulombs, so we're finding the number of electrons that must be removed to leave a positive charge of 1.6×10^{-8} C on the balloon. Thus, charge per electron times number of electrons equals the balloon's charge, and:

$$N = Q_{total}/Q$$

$$N = \frac{1.6 \times 10^{-8}}{1.6 \times 10^{-19}} = 1 \times 10^{11}.$$

33. **(D)** Set the centripetal force equal to the apparent weight on the space station.

$$Mv^2/R = Mg$$

Thus, $g = v^2/R = (55 \text{ m/s})^2/450 \text{ m}.$

Since you have no calculator, do some cancelling with the factors before estimating the nearest answer, which is 6.7 m/s^2.

34. **(D)** The velocity is zero when the spring is at maximum amplitude, or when energy is stored as potential energy. The acceleration is zero at equilibrium, when there is no net force on the spring. The answer combination that satisfies these conditions is **(D)**, where the velocity is zero at 2.0 m, which is the amplitude, and the acceleration is zero at 0, which is equilibrium.

35. **(A)** The beat frequency of 4 Hz indicates a difference of 4 Hz between the saxophone and the trombone. Since the trombone is moving toward the listener, the apparent frequency of the trombone is higher, due to the Doppler effect. Therefore, you hear 444 Hz from the trombone.

AP Physics

ANSWERS AND EXPLANATIONS TO PRACTICE TEST I

36. **(B)** Velocity equals frequency times wavelength. The velocity of sound at room temperature is about 340 m/s—slightly higher with the higher temperature. Thus, the wavelength must be slightly over 1 meter. If you remember the formula V_{sound} = 331 m/s + 0.6 T_C, you find the velocity to be 346 m/s, which gives an even closer approximation to the answer, 1.01 m.

37. **(D)**

$$T_{pendulum} = 2\pi\left(\sqrt{\frac{L}{g}}\right)$$
$$g = 4\pi^2 L/T^2 = = 4\pi^2(1.5)/(3.3)^2$$

An estimate can be made by cancelling the 3.3^2 with π^2, to get about 6.

38. **(A)** In a closed pipe, a standing wave must have a node at the closed end and an antinode at the open end. The wavelength of the fundamental is 4 times the tube length, the wavelength of the first overtone is $\frac{4}{3}$ times the tube length, and the wavelength of the second overtone is $\frac{4}{5}$ times the tube length. Since velocity is constant and $v = f\lambda$, frequency for the second overtone must be 5 times the fundamental frequency.

39. **(C)**

$$T = 2\pi\sqrt{\frac{L}{g}}$$
$$g = 4\pi^2 L/T^2 = 4\pi^2(1 \text{ m})/9$$
$$g = \text{about } 4 \text{ m/s}^2.$$

40. **(B)** Use the concave mirror equation to find image distance:

$$1/f = 1/d_i + 1/d_o$$
$$1/10 = 1/d_i + 1/20$$
$$d_i = 20 \text{ cm}$$

Now use the relationship for magnification:

$$|M| = d_i/d_o = h_i/h_o$$

Since the object and image distances are equal, magnification is one, and the image and object are the same height.

KAPLAN 293

AP Physics

ANSWERS AND EXPLANATIONS TO PRACTICE TEST I

41. **(E)** By the equation $f = \dfrac{1}{2\pi}\sqrt{\dfrac{k}{m}}$, cutting the mass in half would, mathematically, increase f by a factor of square root of two.

42. **(C)** Capacitance is the proportionality constant between charge and charging voltage, $Q = CV$. Thus, $C = Q/V$, and the answer is **(C)**.

43. **(E)** First, the absolute potential, V, due to the first charge is:

    $V = kQ/R$

    $V = (9 \times 10^9 \text{ N·m}^2/\text{C}^2)(3 \times 10^{-6} \text{ C})/(.025 \text{ m})$

    When working without a calculator, think of the denominator as $\dfrac{1}{10}$ times $\dfrac{1}{4}$ and simply put both 10 and 4 as factors in the numerator. As a result, we have:

    $V = (9)(3)(40) \times 10^3 = 1{,}080 \times 10^3 \text{ V}$

    Now, the potential energy of the second charge is equal to the work required to move it into the position where the first potential exists, which is qV.

    $U = qV = (3 \times 10^{-6})(1{,}080 \times 10^3 \text{ V}) = 3{,}240 \times 10^{-3}$ or 3.2 J.

44. **(D)** The force of the +8μC charge on the –5μC charge is:

    $F_E = kQ_1Q_2/R^2$

    $F_1 = (9 \times 10^9 \text{ N·m}^2/\text{C}^2)(+8 \times 10^{-6})(-5 \times 10^{-6})/(2 \text{ m})^2 =$

    $(9)(8)(5) \times 10^{-3} / 2^2$ or 90 mN to the left.

    The force of the –3μC charge on the –5μC charge:

    $F_2 = (9 \times 10^9 \text{ N·m}^2/\text{C}^2)(-3 \times 10^{-6})(-5 \times 10^{-6})/(1 \text{ m})^2$

    $(9)(3)(5) \times 10^{-3}$ or 135 mN to the right

    Combining the two opposite vectors, using right as positive and left as negative:

    $F_{net} = F_2 - F_1 = 45$ mN to the right.

45. **(D)** Both Coulomb's law and Newton's Universal Law of Gravitation describe forces that decrease with the inverse square of the distance and act at a distance. However, only Coulomb forces may be repulsive.

AP Physics

ANSWERS AND EXPLANATIONS TO PRACTICE TEST I

46. **(A)** Using $E = kQ_1Q_2/R^2$, with the field outward from the positive charge and inward toward the negative charge, we can see that the vectors will never cancel each other to produce zero between the charges. Only to the left of the smaller charge is there a possibility of a zero field. The field to the left from the $+2q$ and field to the right from the more distant but larger $-4q$ could cancel to produce zero.

47. **(D)** First, remember that moving across field lines would be an "equipotential," with no change in potential in this direction. By $V = Ed$, and using the distance *along* the field defined by ABcos 30 °,

$$V = (100 \text{ N/C})(0.2)\left(\frac{\sqrt{3}}{2}\right) = 10\sqrt{3}$$

Estimating the answer, we choose **(D)**, which is 17.3 V.

48. **(D)** Work is equal to change in electric potential energy, qV.

$$W = (3 \text{ C})(1.5 \text{ V}) = 4.5 \text{ J.}$$

49. **(E)** Using the right hand rule and then reversing direction for the negative charge (or using the left hand in the first place), we find a magnetic force directed upward on the page. The electron will move in a counter-clockwise circular motion. $\mathbf{F} = q\mathbf{v}\times\mathbf{B}$.

50. **(D)** Lenz's Law describes the production of a magnetic field that produces a force that opposes the motion. A metal bar moving across a magnetic field will itself produce a magnetic field that opposes the motion and thus slows it down.

51. **(A)** By our common usage and definition, the north pole of a magnetic compass points toward the Magnetic North Pole of Earth. However, it should be noted that since opposite poles attract, the defined north pole of a magnetic compass must point toward a south magnetic pole. Thus, what we call Earth's North Magnetic Pole is actually a south magnetic pole, and what we call Earth's South Magnetic Pole must be a north magnetic pole.

52. **(B)** Only the velocity component perpendicular to the magnetic field lines will be affected by the magnetic field. Use $(v \sin 60°)$ for the velocity component and the right hand rule to determine direction.

$$\mathbf{F} = q\mathbf{v}\times\mathbf{B} = (1.6 \times 10^{-19})(2.0 \times 10^6)(\sin 60°)(1.5 \text{ T}) = 4.2 \times 10^{-13} \text{ N into the page}$$

53. **(E)** The magnetic field in Teslas is equivalent to $\dfrac{\text{N}}{\text{A·m}}$. (This can be derived from the equation $|F| = qvB$.) Only choice **(E)** is not equivalent.

KAPLAN 295

AP Physics
ANSWERS AND EXPLANATIONS TO PRACTICE TEST I

54. **(B)** The current I_1 will produce a magnetic field into the page at B, and the current I_2 will produce a magnetic field into the page at B. The two fields will repel, causing the wires to repel.

55. **(C)** The magnetic fields due to the currents in the wires will be into the page at every point between the wires, so they will not cancel at any point between the wires.

56. **(B)** The metal piece will be attracted to the end of the bar magnet.

57. **(D)** Using the right hand rule, the particle experiences a force to the right and will thus curve to the right. $\mathbf{F} = q\,\mathbf{v} \times \mathbf{B}$.

58. **(C)** The field between the charges cannot be zero, because the vectors are the same direction (toward the negative and away from the positive) and can never cancel. However, along the x-axis to the right of the negative charge, it is possible for the electric field vectors to cancel. Since the positive charge is larger in magnitude, the magnitudes of the fields could be equal at a point farther from the positive and closer to the negative, and in this region the field from the positive is to the right and field from the negative is to the left.

59. **(A)** Two 20-ohm resistors in parallel would have an equivalent resistance:

    $$R_{eq} = (20)^2/(40), \text{ or } 10 \text{ ohms.}$$

    (Remember: This "product over sum" rule only works for two resistors in parallel.)

    Using Ohm's Law:

    $$I = V/R, \text{ or } 1 \text{ amp.}$$

    Since the two resistors in parallel are of equal value, the current will split evenly in both branches so that 0.5 amp flows through each resistor.

60. **(A)** Fusion releases more energy than fission, per nucleon. However, comparing a nucleus of Uranium (235 nucleons), to two nuclei of Deuterium (4 nucleous), the Uranium produces more energy (about 10 times as much).

61. **(C)** In beta [plus] or positron decay, a proton in the nucleus in converted to a neutron, lowering the atomic number by one but keeping the mass number constant.

62. **(D)** Nuclei will fuse, but only at very high temperatures, where they move at high enough speeds to achieve fusion.

296 KAPLAN

AP Physics
ANSWERS AND EXPLANATIONS TO PRACTICE TEST I

63. **(C)** A beta⁻, or high energy electron, would be produced from a nuclear neutron decay that also produced a proton plus a neutrino.

64. **(E)** Use $E = mc^2$, with m equal to about 0.2 μ.

From the Table of Information, we find that 1 μ = 931 MeV/c^2, so we estimate 0.2 μ as 180 or 190 MeV/c^2.

Substituting: $E = mc^2 = (180 \text{ MeV}/c^2)c^2 = 180$ MeV.

Here's a case where using units is essential to finding the answer!

65. **(C)** Half life is the time for half of the nuclei in a given mass to decay, but one given nucleus is either in that half the number that decay or it is not, giving it a 50 : 50 chance during that time to decay.

66. **(E)** All other quantities are conserved in nuclear reactions except velocity. For example, fission might send daughter nuclei in various directions from a stationary nucleus, so that momentum is still conserved, but velocity is not conserved.

67. **(B)** Increasing the current in the bottom loop would increase the magnetic field upward through the loop, using the right hand rule. By Lenz's Law, this increase in magnetic field upward through the top loop will induce an EMF in the top loop opposing the change. Thus, the current in the top loop will decrease in response.

68. **(C)** Force is measured in Newtons in the SI system.

69. **(E)** Both atomic number (the number on the bottom left) and mass number (the upper number) must balance on the two sides of the equation. In choice **(E)**, knowing that a neutrino has atomic number 0 and mass number 0, the atomic number sum is 82 on the left side and 84 on the right side, which does not balance.

70. **(A)** In a photoelectric experiment, the frequency must be above a specific threshold frequency in order to produce an current at all. Making the light "bluer" would increase its frequency.

$$hf = \phi + K$$

KAPLAN 297

SECTION II: FREE RESPONSE

1. (15 points)

    Consider the pulley apparatus below. A box of mass 10 kg is being pulled across a table by a falling box of mass 7.5 kg that is suspended by a cord over a frictionless, massless pulley. The coefficient of kinetic friction between the box and table is 0.20.

    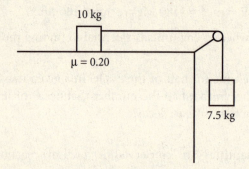

    (a) Find the acceleration of the two-block system. (3 points)

    Answer:

    The net force on the system equals the downward force of gravity on the 7.5 kg block (73.5 N) minus friction ($F_f = \mu mg = 0.2(10)(9.8) = 19.6$ N) on the 10 kg block.

    Thus

    $$F_{net} = 73.5 - 19.6 = 53.9 \text{ N.}$$
    $$\text{Then acceleration equals F/m: } a = \frac{F}{m} = \frac{53.9 \text{ N}}{17.5 \text{ kg}} = 3.08 \text{ m/s}^2.$$

    (b) What is the tension in the cord? (3 points)

    Answer:

    The tension minus friction has to equal the net force on the 10 kg block. We know the friction force equals 19.6 N, and given the acceleration of 3.08 m/s², we know that the net force must be 30.8 N on the 10 kg block, by F = ma. Thus we have:

    $$T - 19.6 \text{ N} = 30.8 \text{ N}$$

    which means that T = 50.4 N.

ANSWERS AND EXPLANATIONS TO PRACTICE TEST I

(c) Now suppose the hanging mass is replaced by a downward force of 75 N. What will be the resulting acceleration of the system? (3 points)

Answer:

The same process as part (a) will hold except that the downward force will now equal 75 N and the total mass of the system now equals 10 kg (no second mass to accelerate). Friction remains constant, so the net force =

$$75 \text{ N} - 19.6 \text{ N} = 55.4 \text{ N.}$$

Acceleration, then, equals

$$F/m = 55.4 \text{ N}/10 \text{ kg} = 5.54 \text{ m/s}^2.$$

(d) What is the new tension? (3 points)

Answer:

Now, the only force acting on the block besides friction is the tension, and it comes from the 75 N downward force. Thus, the new tension is 75 N. Combined with the friction force of 19.6 N in the other direction, the net force of 55.4 N is easy to find.

(e) How much mass should be added to the first block for it to slide at constant velocity? (3 points)

Answer:

For the block not to accelerate, the net force on it must be zero. There are two forces acting on it in the horizontal direction: friction and tension. The 75 N tension force won't change, so we need to find the mass required for friction to equal 75 N.

$$F_f = 75 \text{ N} = \mu mg = 0.2(x)(9.8)$$

Solving this for our unknown mass (x), we get a total mass of 38.3 kg. Since we already have 10 kg, we will need an additional 28.3 kg.

2. (15 points)

A block slides along a frictionless surface toward a second block attached to the end of a spring as shown below. When the blocks collide they stick together.

$V = 8$ m/s $v = 0$ m/s $k = 250$ N/m
$m = 5$ kg $m = 2$ kg

KAPLAN 299

AP Physics
ANSWERS AND EXPLANATIONS TO PRACTICE TEST I

(a) What is the speed of the two blocks immediately after the collision? (5 points)

Answer:

Momentum must be conserved. The initial momentum is (8kg)(5 m/s), or 40 kg·m/s. After the collision, momentum must still be 40 kg·m/s, but the new mass is 7 kg.

Thus:

$$p = mv$$

$$40 = 7(v), \text{ and } v = 5.7 \text{ m/s.}$$

(b) What impulse is given to the first block during the collision? (5 points)

Answer:

Impulse is simply change in momentum $= m(v_f - v_o)$

$$m\Delta v = 5\text{kg} \, (-2.3 \text{ m/s}) = -11.5 \text{ kg·m/s.}$$

Note: Units for this answer could also be Newton-seconds.

(c) How far will the spring compress? (3 points)

Answer:

Total energy must be conserved, so the kinetic energy of the blocks after the collision is equal to the potential energy of the spring system once it is compressed.

$$\frac{1}{2} mv^2 = \frac{1}{2} kA^2, \text{ where A is the amplitude, or maximum compression}$$

$$(0.5)(7 \text{ kg}) (5.7 \text{ m/s})^2 = (0.5)(250 \text{ N/m})(A)^2$$

$$A = 0.95 \text{ m}$$

(d) What is the maximum speed the blocks will attain once they begin oscillating on the spring? (2 points)

Answer:

Since energy is conserved, the blocks exchange energy between potential and kinetic. Thus, the maximum speed of the blocks will be same as the speed they had when they hit the spring, i.e., 5.7 m/s.

AP Physics

ANSWERS AND EXPLANATIONS TO PRACTICE TEST I

3. (15 points)

A 50 g piece of ice at an initial temperature that is unknown is placed into a glass containing 400 g of water at 20°C. The glass and ice come to thermal equilibrium at 5°C.

(a) (i) If the ice floats with 90 percent of its volume under water, what is its specific gravity? (2 points)

Answer:

Since the ice need only displace 90 percent of its volume to displace an amount of water equal to its own weight, the ice must only have a density that is 90 percent that of water. Thus, the specific gravity, which is density of ice divided by density of water, must be 0.9.

(ii) What is the buoyant force on the floating ice? (3 points)

Answer:

If the ice is floating, the buoyant force on it is equal to the weight of the ice.

$$W_{ice} = m_{ice}g = (0.050 \text{ kg})(9.8 \text{ m/s}^2) = 0.49 \text{ N}$$

(b) The ice and water mixture comes to equilibrium at 5°C. What was the original temperature of the ice? (5 points)

Answer:

If the ice melts and the ice-water mixture comes to equilibrium, energy must be conserved, so:

Loss of heat by ice + Gain in heat by water = 0

The total heat lost by the ice occurs in three stages: heating the ice from its original temperature to its melting point, changing the ice to water, and heating the melted water to equilibrium temperature:

$$Q_{warming\ ice} + Q_{melting\ ice} + Q_{heating\ ice\ water} + Q_{cooling\ warm\ water} = 0$$
$$m_{ice}c_{ice}\Delta T + m_{ice}L_{ice} + m_{icewater}c_{water}\Delta T + m_{water}c_{water}\Delta T = 0$$
$$(0.050 \text{ kg})(2,000 \text{ J/kgC°})(0°C - T_o) + (0.050 \text{ kg})(3.35 \times 10^5 \text{ J/kg}) +$$
$$(0.050 \text{ kg})(4,180 \text{ J/kgC°})(5°C) + (.400 \text{ kg})(4,180 \text{ J/kgC°})(5°C - 20°C) = 0$$

Solving for the original temperature of ice, T_o:

$$-100 \text{ } T_o + (1.675 \times 10^4) + 1,045 + 6,270 - (2.508 \times 10^4) = 0$$
$$T_o = -10.2°C.$$

KAPLAN 301

AP Physics

ANSWERS AND EXPLANATIONS TO PRACTICE TEST I

(c) Now, 0.27 kg from a large block of the original ice melts as a result of a 0.75 kg chunk of hot metal being placed on it. (5 points)

Answer:

First, we must realize that the size of the original block of ice does not matter in this problem. All we are concerned with is the mass of ice that melts. We obtain the original temperature of the ice from our answer to part (b). Also, since the mass of the ice is so large, we assume that the final temperature of the metal is that of the melted ice on which it sits, 0°C.

As before, energy is conserved:

$$Q_{ice} + Q_{melt\ ice} + Q_{metal} = 0$$

$$m_{ice}c_{ice}\Delta T_{ice} + m_{ice}L_{ice} + m_{metal}c_{metal}\Delta T_{metal} = 0$$

$$(0.27\ kg)(2,000\ J/kg°C)(10.2°C.) + (0.27\ kg)(3.35 \times 10^5\ J/kg) +$$

$$(0.75\ kg)\ (c)(-200°C) = 0$$

Solving for specific heat of metal: c = 640 J/kg°C (5 points)

4. **(15 points)**

A musician blows into an open pipe and produces its fundamental frequency. The pipe is 0.70 meters long. (Use v_{sound} = 343 m/s.)

(a) What is the fundamental frequency? (5 points)

Answer:

In an open pipe, the fundamental frequency has a wavelength that is exactly twice the length of the pipe (since one-half of a wavelength fits into the pipe). Thus, the wavelength of this sound is 1.4 m, and the frequency is:

$$f = \frac{v}{\lambda} = \frac{343\ ^m/_s}{1.4\ m} = 245\ Hz$$

(b) If he wanted to play a higher note, what would be the next higher frequency that can be played on this pipe? (4 points)

Answer:

The next frequency that will resonate is that which has a wavelength that completely fits inside the pipe. This means that the wavelength is 0.7 meters, or exactly half the original wavelength. When the wavelength is cut in half, the frequency doubles (assuming the speed of the sound wave remains constant—a safe assumption in this case). Thus the next higher frequency is 490 Hz.

(c) How many wavelengths fit inside the tube in this case? (2 points)

Answer:

As explained above, one complete wavelength fits inside the pipe.

(d) If the musician were to replace his open pipe with a closed pipe, what is one possible length for the pipe if he wants to produce the same (higher) frequency? (4 points)

Answer:

To produce a note with a frequency of 490 Hz in a closed pipe, the shortest pipe would have a length of $\lambda/4$. The wavelength is still 0.7 meters, so the pipe would have to be 0.7/4 meters long, or 0.175 meters.

Another pipe that would resonate this note would have to be 3 times longer, or 0.525 meters. Subsequent lengths would be odd multiples of the original length (5L, 7L, 9L,...).

5. (15 points)

(a) Using the diagram below, locate the image formed by drawing a ray diagram. (6 points)

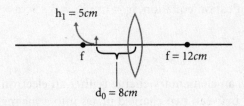

Answer:

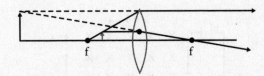

(b) Calculate the image distance. (3 points)

Answer:

Using the lens equation:

$$1/d_0 + 1/d_i = 1/f.$$

You get:

$$1/8 + 1/d_i = \frac{1}{12}$$

Solving for d_i:

$$d_i = -24 \text{ cm}.$$

(c) What kind of image is this? Clearly explain how you know this. (2 points)

Answer:

The image is virtual for these reasons: it is upright, behind the lens, has a negative image distance, and is formed from projected rays (dashed lines).

(d) Calculate the magnification of this lens. (2 points)

Answer:

$$m = h_i/h_0 = -d_i/d_0 = -(-24)/8 = 3$$

(e) Calculate the height of the image formed. (2 points)

Answer:

From the magnification equation, $h_i = m \cdot h_0 = 3 \cdot 5\text{cm} = 15$ cm.

6. (15 points)

In the diagram below, an elementary charge (either an electron or a proton, you must decide) is accelerated between two charged plates into a magnetic field. The mass of a proton = $1.67 \cdot 10^{-27}$ kg. The mass of an electron = $9.11 \cdot 10^{-31}$ kg.

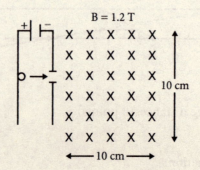

ANSWERS AND EXPLANATIONS TO PRACTICE TEST I

AP Physics

(a)   What is the sign of the elementary charge? (2 points)

Answer:

Since the left plate is positively charged (note the battery diagram), the elementary charge must also be positive, which makes it a proton.

(b)   If the charge is traveling at a speed of $2.0 \cdot 10^5$ m/s when it passes through the opening, what is the voltage charging the two plates? (4 points)

Answer:

The potential energy of the proton = qV. That energy is then converted to kinetic energy. Thus,

$$\frac{1}{2}mv^2 = qV$$

$$\left(\frac{1}{2}\right)mv^2 = (0.5)(1.67 \times 10^{-27} \text{ kg}) (2.0 \times 10^5 \text{ m/s})^2 = 3.34 \times 10^{-17} \text{ J}$$

$$qV = (1.6 \times 10^{-19} \text{ C})(V) = 3.34 \times 10^{-17} \text{ J}$$

So,

$$V = 209 \text{ V}.$$

(c)   Describe the path of the charge in the magnetic field. (4 points)

Answer:

The proton enters the magnetic field perpendicular to the field, so it will follow a circular path through the field. The path will be counter-clockwise, according to the right-hand rule.

(d)   Find the distance from the entrance point to where the charge leaves the magnetic field. (5 points)

Answer:

When a charged particle travels in a circular path through a magnetic field, the magnetic force provides the centripetal force:

$$mv^2/r = qvB$$

$$r = mv/Bq = [(1.67 \times 10^{-27} \text{ kg})(2.0 \times 10^5 \text{ m/s})]/[(1.2 \text{ T})(1.6 \times 10^{-19} \text{ C})] = 0.0017 \text{ m}.$$

This is the radius of the circle, so when the proton leaves the magnetic field, it will be one diameter away from where it entered. Thus the answer is 0.0034m, or 3.4 mm from its entrance point.

KAPLAN   305

PRACTICE TEST II
ANSWER SHEET

1 Ⓐ Ⓑ Ⓒ Ⓓ Ⓔ

2 Ⓐ Ⓑ Ⓒ Ⓓ Ⓔ

3 Ⓐ Ⓑ Ⓒ Ⓓ Ⓔ

4 Ⓐ Ⓑ Ⓒ Ⓓ Ⓔ

5 Ⓐ Ⓑ Ⓒ Ⓓ Ⓔ

6 Ⓐ Ⓑ Ⓒ Ⓓ Ⓔ

7 Ⓐ Ⓑ Ⓒ Ⓓ Ⓔ

8 Ⓐ Ⓑ Ⓒ Ⓓ Ⓔ

9 Ⓐ Ⓑ Ⓒ Ⓓ Ⓔ

10 Ⓐ Ⓑ Ⓒ Ⓓ Ⓔ

11 Ⓐ Ⓑ Ⓒ Ⓓ Ⓔ

12 Ⓐ Ⓑ Ⓒ Ⓓ Ⓔ

13 Ⓐ Ⓑ Ⓒ Ⓓ Ⓔ

14 Ⓐ Ⓑ Ⓒ Ⓓ Ⓔ

15 Ⓐ Ⓑ Ⓒ Ⓓ Ⓔ

16 Ⓐ Ⓑ Ⓒ Ⓓ Ⓔ

17 Ⓐ Ⓑ Ⓒ Ⓓ Ⓔ

18 Ⓐ Ⓑ Ⓒ Ⓓ Ⓔ

19 Ⓐ Ⓑ Ⓒ Ⓓ Ⓔ

20 Ⓐ Ⓑ Ⓒ Ⓓ Ⓔ

21 Ⓐ Ⓑ Ⓒ Ⓓ Ⓔ

22 Ⓐ Ⓑ Ⓒ Ⓓ Ⓔ

23 Ⓐ Ⓑ Ⓒ Ⓓ Ⓔ

24 Ⓐ Ⓑ Ⓒ Ⓓ Ⓔ

25 Ⓐ Ⓑ Ⓒ Ⓓ Ⓔ

26 Ⓐ Ⓑ Ⓒ Ⓓ Ⓔ

27 Ⓐ Ⓑ Ⓒ Ⓓ Ⓔ

28 Ⓐ Ⓑ Ⓒ Ⓓ Ⓔ

29 Ⓐ Ⓑ Ⓒ Ⓓ Ⓔ

30 Ⓐ Ⓑ Ⓒ Ⓓ Ⓔ

31 Ⓐ Ⓑ Ⓒ Ⓓ Ⓔ

32 Ⓐ Ⓑ Ⓒ Ⓓ Ⓔ

33 Ⓐ Ⓑ Ⓒ Ⓓ Ⓔ

34 Ⓐ Ⓑ Ⓒ Ⓓ Ⓔ

35 Ⓐ Ⓑ Ⓒ Ⓓ Ⓔ

36 Ⓐ Ⓑ Ⓒ Ⓓ Ⓔ

37 Ⓐ Ⓑ Ⓒ Ⓓ Ⓔ

38 Ⓐ Ⓑ Ⓒ Ⓓ Ⓔ

39 Ⓐ Ⓑ Ⓒ Ⓓ Ⓔ

40 Ⓐ Ⓑ Ⓒ Ⓓ Ⓔ

41 Ⓐ Ⓑ Ⓒ Ⓓ Ⓔ

42 Ⓐ Ⓑ Ⓒ Ⓓ Ⓔ

43 Ⓐ Ⓑ Ⓒ Ⓓ Ⓔ

44 Ⓐ Ⓑ Ⓒ Ⓓ Ⓔ

45 Ⓐ Ⓑ Ⓒ Ⓓ Ⓔ

46 Ⓐ Ⓑ Ⓒ Ⓓ Ⓔ

47 Ⓐ Ⓑ Ⓒ Ⓓ Ⓔ

48 Ⓐ Ⓑ Ⓒ Ⓓ Ⓔ

49 Ⓐ Ⓑ Ⓒ Ⓓ Ⓔ

50 Ⓐ Ⓑ Ⓒ Ⓓ Ⓔ

51 Ⓐ Ⓑ Ⓒ Ⓓ Ⓔ

52 Ⓐ Ⓑ Ⓒ Ⓓ Ⓔ

53 Ⓐ Ⓑ Ⓒ Ⓓ Ⓔ

54 Ⓐ Ⓑ Ⓒ Ⓓ Ⓔ

55 Ⓐ Ⓑ Ⓒ Ⓓ Ⓔ

56 Ⓐ Ⓑ Ⓒ Ⓓ Ⓔ

57 Ⓐ Ⓑ Ⓒ Ⓓ Ⓔ

58 Ⓐ Ⓑ Ⓒ Ⓓ Ⓔ

59 Ⓐ Ⓑ Ⓒ Ⓓ Ⓔ

60 Ⓐ Ⓑ Ⓒ Ⓓ Ⓔ

61 Ⓐ Ⓑ Ⓒ Ⓓ Ⓔ

62 Ⓐ Ⓑ Ⓒ Ⓓ Ⓔ

63 Ⓐ Ⓑ Ⓒ Ⓓ Ⓔ

64 Ⓐ Ⓑ Ⓒ Ⓓ Ⓔ

65 Ⓐ Ⓑ Ⓒ Ⓓ Ⓔ

66 Ⓐ Ⓑ Ⓒ Ⓓ Ⓔ

67 Ⓐ Ⓑ Ⓒ Ⓓ Ⓔ

68 Ⓐ Ⓑ Ⓒ Ⓓ Ⓔ

69 Ⓐ Ⓑ Ⓒ Ⓓ Ⓔ

70 Ⓐ Ⓑ Ⓒ Ⓓ Ⓔ

PRACTICE TEST II
AP PHYSICS B

Section I—Multiple Choice

Time: 90 minutes
Questions: 70
Points: 90

Directions: Each of the questions below is followed by five answer choices. Select the best answer, and fill in the corresponding oval on your answer sheet. This portion of the test should be taken without a calculator or reference to formulas.

Note: 10 m/s^2 my be used for g in calculations.

1. On a displacement-time graph, a horizontal straight line corresponds to motion at:

   (A) zero speed
   (B) constant, non-zero speed
   (C) increasing speed
   (D) decreasing speed
   (E) decreasing acceleration

2. Ball A is thrown horizontally and ball B is dropped vertically from the same height at the same moment. Which of the following is true?

   (A) Ball A reaches the ground first.
   (B) Ball B reaches the ground first.
   (C) Ball A has the greater speed when it reaches the ground.
   (D) Ball B has the greater speed when it reaches the ground.
   (E) Both balls have the same speed when they reach the ground.

3. An airplane is flying horizontally at an altitude of 490 meters when a wheel falls from it. If there were no air resistance, the wheel would strike the ground in:

   (A) 10 s
   (B) 30 s
   (C) 50 s
   (D) 80 s
   (E) 100 s

4. A stone is dropped from a cliff. After it has fallen 30 m, its speed is:

   (A) 17 m/s
   (B) 24 m/s
   (C) 44 m/s
   (D) 74 m/a
   (E) 588 m/s

5. A horizontal force of 40 N is needed to start a 10 kg steel box moving across a wooden floor. The coefficient of static friction is:

   (A) 0.08
   (B) 0.25
   (C) 0.4
   (D) 0.8
   (E) 2.5

GO ON TO THE NEXT PAGE. ➡

KAPLAN   307

AP Physics
PRACTICE TEST II

6. An object traveling in a circle at constant speed:
   (A) has a constant velocity
   (B) is not in accelerated motion
   (C) has an inward, radial acceleration
   (D) has an outward, radial acceleration
   (E) has a constant tangential acceleration

7. The radius of the path of an object in uniform circular motion is doubled. For its speed to remain the same, the centripetal force on the object must be:
   (A) one-fourth as much as before
   (B) half as great as before
   (C) the same as before
   (D) twice as great as before
   (E) four times as great as before

8. The moon's mass is about one-sixth of the earth's mass. Compared to the gravitational force the earth exerts on the moon, the gravitational force the moon exerts on the earth:
   (A) is one-sixth as much
   (B) is one-half as much
   (C) is the same
   (D) is twice as much
   (E) is six times as much

9. A car is traveling at 10 m/s on a road such that the coefficient of friction between its tires and the road is 0.5. The minimum turning radius of the car is:
   (A) 5.0 m
   (B) 10 m
   (C) 20 m
   (D) 40 m
   (E) 50 m

Questions 10-14 refer to the following graph.

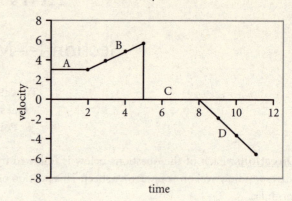

10. During which labeled segment on the graph is the object not moving at all?
    (A) A
    (B) B
    (C) C
    (D) D
    (E) There is no section on the graph where the object is at rest.

11. During which labeled segment on the graph is the object undergoing the largest magnitude of acceleration?
    (A) A
    (B) B
    (C) C
    (D) D
    (E) There is no section on the graph where the object is accelerating.

12. During which labeled segment on the graph is the object moving but not accelerating?
    (A) A
    (B) B
    (C) C
    (D) D
    (E) The object is accelerating in more than one section.

GO ON TO THE NEXT PAGE.

13. If the velocity is measured in meters per second, at what time does the object have a speed of 4 m/s?

    (A) 1 s
    (B) 4 s
    (C) 5 s
    (D) 3 s and 10 s
    (E) 8 s

14. How far had the object traveled after the first 5 seconds?

    (A) 5 m
    (B) 6.5 m
    (C) 19.5 m
    (D) 22.5 m
    (E) 24 m

15. Consider two satellites in circular orbits around the Earth but at different distances from the Earth.

    (A) Both experience the same centripetal acceleration.
    (B) The object nearer to Earth experiences the greater centripetal acceleration.
    (C) The object farther from Earth experiences the greater centripetal acceleration.
    (D) It depends upon the masses of the satellites, which are not known.
    (E) Neither has centripetal acceleration.

16. Which of the following would make an object of given mass, m, appear to increase in weight when compared to its weight when stationary on the surface of the earth?

    I. object accelerating upward in an elevator
    II. object in free fall
    III. object in the Space Shuttle as it takes off from Earth's surface

    (A) I only
    (B) II only
    (C) III only
    (D) I and III
    (E) I and II

17. Which of the following is a correct free body diagram for a box sitting on a ramp?

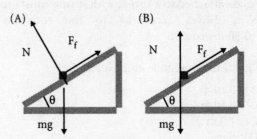

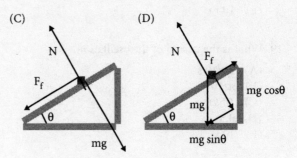

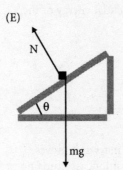

GO ON TO THE NEXT PAGE.

Questions 18-21 apply to the following situation:

A 1 kg mass attached to a spring with spring constant 100 N/m moves according to the equation $x(t) = (0.50 \text{ m}) \cos 10t$.

18. What is the amplitude of the oscillation?

    (A) 0.10 m
    (B) 0.50 m
    (C) 1.0 m
    (D) 5 m
    (E) 10 m

19. What is the period of the oscillation?

    (A) $2\pi/10$ s
    (B) $\pi/10$ s
    (C) $10/2\pi$ s
    (D) 0.5 s
    (E) 10 s

20. What is the maximum potential energy of the spring-mass system?

    (A) 100 J
    (B) 50 J
    (C) 25 J
    (D) 12.5 J
    (E) 5.0 J

21. What is the velocity of the mass as it moves through its equilibrium position, assuming no friction?

    (A) 10 m/s
    (B) 2π m/s
    (C) 5 m/s
    (D) 2.5 m/s
    (E) 0.5 m/s

Questions 22-25: Use the graphical relationship shown here to answer the following questions.

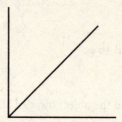

22. If the graph is force vs time, the area under the graph would represent:

    (A) change in velocity
    (B) change in momentum
    (C) change in acceleration
    (D) work
    (E) change in kinetic energy

23. If the graph is position vs time, it could represent:

    (A) motion at constant velocity
    (B) motion with increasing velocity
    (C) motion with increasing acceleration
    (D) motion with decreasing acceleration
    (E) no motion at all

24. The graph could represent:

    (A) velocity of an oscillating pendulum as a function of time
    (B) acceleration of an oscillating pendulum as a function of time
    (C) position of a vibrating mass on a spring as a function of time
    (D) velocity of a falling object as a function of time
    (E) displacement of a falling object as a function of time

25. If the graph is a plot of energy as a function of frequency of light, the slope would be the value of:

    (A) the work function
    (B) Planck's constant
    (C) the speed of light
    (D) the relativistic momentum
    (E) wavelength

GO ON TO THE NEXT PAGE.

AP Physics
PRACTICE TEST II

26. Rank the following forms of electromagnetic radiation by their wavelengths, from shortest to longest:

    1: visible light
    2: radio waves
    3: Gamma rays
    4: X-rays

    (A) 2,1,3,4
    (B) 3,4,1,2
    (C) 2,1,4,3
    (D) 1,3,4,2
    (E) 4,3,2,1

27. A beam of laser light in a room (full of air) is pointed at a diamond, index of refraction n = 1.4. What is the critical angle for the light entering the diamond?

    (A) 48.6°
    (B) 2.42°
    (C) 24.4°
    (D) There is no critical angle for air into a diamond.
    (E) 0.41°

28. If you shine white light through a diffraction grating with 1,300 lines per inch, which color will have its first order maximum farthest from the light's original path?

    (A) blue
    (B) green
    (C) yellow
    (D) orange
    (E) red

29. Which of the following is an application of concave mirror reflections?

    (A) the word AMBULANCE written backwards on the front of ambulances
    (B) satellite dishes
    (C) convenience store mirrors
    (D) automobile mirrors on the passenger side that say "Objects in mirror are closer than they appear"
    (E) rear-view mirrors in a car that can be tilted at night to make headlights less bright

30. When a large soap bubble is viewed in white light, it produces many colors. What is the best explanation for this phenomenon?

    (A) There is no explanation for this phenomenon.
    (B) Different thicknesses in the bubble reflect different colors.
    (C) The bubble acts as a prism and separates the light into the spectrum of colors.
    (D) White light is made up of all the colors of the rainbow.
    (E) You can view it from any angle, so you see all the colors.

31. Why does a sound wave diffract much more significantly than a light wave when each passes through a typical classroom door?

    (A) Sound waves carry more energy than light waves.
    (B) Sound waves travel more slowly.
    (C) Sound waves are longitudinal.
    (D) Sound waves have a smaller frequency than light waves.
    (E) They both diffract equally, as long as the door opening remains the same.

32. As you increase the wavelength of light shining through a single slit aperture, what happens to the interference pattern formed?

    (A) nothing
    (B) It becomes more spread out.
    (C) It becomes less spread out.
    (D) It depends on the size of the opening.
    (E) Single slits don't produce interference patterns.

GO ON TO THE NEXT PAGE. ➡

KAPLAN 311

33. Locate the image and find its height for the following situation: a 10 cm tall object is placed 15 cm in front of a convex lens whose focal length is 20 cm.

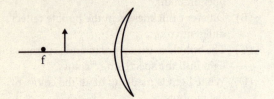

	Location	Height
(A)	60 cm	2.5 cm
(B)	-60 cm	40 cm
(C)	20 cm	5 cm
(D)	-30 cm	2.5 cm
(E)	-20 cm	10 cm

34. An object placed "d_0" meters in front of a convex mirror with focal length "$-f$" produces an image that is $\frac{1}{2}$ the size of the original object. How far from the mirror should the same object be placed to produce an image that is $\frac{1}{4}$ the size of the original object?

   (A) $\frac{1}{2}d_0$
   (B) $2d_0$
   (C) $3f$
   (D) $\frac{1}{4}f$
   (E) $4d_0$

35. Which of the following contributes to the interference patterns formed in thin-film interference?

   I. Thickness of the film
   II. Change in frequency of the light entering the film
   III. Difference in wavelength of light in the two media

   (A) I only
   (B) III only
   (C) I and III
   (D) II and III
   (E) I, II, and III

36. All of the following scenarios are possible EXCEPT:

   (A) A concave mirror produces a real image.
   (B) A plane mirror produces a virtual image.
   (C) A convex mirror produces a virtual image.
   (D) A convex lens produces a real image.
   (E) A concave lens produces a real image.

37. The depth in fresh water at which the total pressure is 3 atmospheres is approximately:

   (A) 5 m
   (B) 10 m
   (C) 15 m
   (D) 20 m
   (E) 30 m

38. A piece of wood sinks in water. The buoyant force on the wood is:

   (A) zero
   (B) less than its weight
   (C) equal to its weight
   (D) more than its weight
   (E) dependent upon the density of the wood but not the size

AP Physics
PRACTICE TEST II

39. In a transverse wave, the individual particles of the medium:

    (A) move in circles
    (B) move in ellipses
    (C) move parallel to the direction of travel of the wave
    (D) move perpendicular to the direction of travel of the wave
    (E) move alternately parallel and perpendicular to the direction of travel

40. The rate at which heat is conducted through a material does NOT depend on:

    (A) temperature difference across the material
    (B) surface area of the material
    (C) thickness of the material
    (D) specific heat of the material
    (E) conductivity of the material

41. Which of the following is NOT an electromagnetic wave?

    (A) radio waves
    (B) blue light
    (C) X-rays
    (D) sound waves
    (E) infrared

42. A police car with its siren sounding is approaching a stationary observer. To the observer, the siren appears to be:

    (A) higher in pitch than it actually is
    (B) lower in pitch than it actually is
    (C) the same pitch as it actually is
    (D) lower in frequency than it actually is
    (E) moving slower than normal

43. As a wave passes through an opening that is approximately equal to its wavelength, the wave will bend. This is a demonstration of:

    (A) diffraction
    (B) diffusion
    (C) deflection
    (D) refraction
    (E) dissociation

44. A tuning fork of frequency 256 Hz is sounded along with a tuning fork of unknown frequency. The observer hears 4 beats per second. The frequency of the second tuning fork is:

    (A) 252 Hz
    (B) 260 Hz
    (C) 4 Hz
    (D) either 252 or 260 Hz
    (E) 1,024 Hz

45. An adiabatic system is one in which:

    (A) no heat enters or leaves the system
    (B) the temperature remains constant
    (C) volume remains constant
    (D) the system does no work, nor is work done on it
    (E) the pressure of the system remains constant

46. Bernoulli's equation is based on:

    (A) the second law of thermodynamics
    (B) Newton's second law of motion
    (C) conservation of charge
    (D) conservation of momentum
    (E) conservation of energy

47. A fluid moving through a section of pipe at velocity v_1 is then sent into a section of pipe with one-half the diameter. The relationship of the velocity v_2 of the fluid in the smaller pipe to the original velocity is:

    (A) $v_2 = v_1$
    (B) $v_2 = v_1/2$
    (C) $v_2 = 2 v_1$
    (D) $v_2 = 4 v_1$
    (E) $v_2 = v_1/4$

GO ON TO THE NEXT PAGE. ➡

KAPLAN 313

AP Physics
PRACTICE TEST II

48. As a ray of light moves from a higher index of refraction to a lower index of refraction:

    (A) velocity decreases, while frequency and wavelength remain the same.
    (B) velocity increases, while frequency and wavelength remain the same.
    (C) velocity and wavelength decrease, while frequency remains the same.
    (D) velocity and wavelength increase, while frequency remains the same.
    (E) velocity and frequency decrease, while wavelength remains the same

49. An ideal gas in a sealed container is heated so that the temperature of the gas increases with no change in volume. Which of the following is true?

    I. pressure of gas increases

    II. work is done on the gas

    III. average kinetic energy of the molecules remains constant

    (A) I only
    (B) II only
    (C) III only
    (D) I, II, and III
    (E) I and II

50. The total fluid pressure on a scuba diver at the bottom of a lake does NOT depend upon:

    (A) atmosperic pressure
    (B) density of the water
    (C) water depth
    (D) surface area of the lake
    (E) value of g at that location

51. An increase in sound decibel level from 30 dB to 40 dB corresponds to :

    (A) an increase in intensity by 100 times
    (B) an increase in intensity by 10 times
    (C) no change in actual sound intensity
    (D) a decrease in sound intensity
    (E) an increase in pitch

52. All of the following are correct statements about electromagnetic waves EXCEPT:

    (A) They all travel at the same speed in a vacuum.
    (B) They can be polarized.
    (C) Frequency is inversely proportional to wavelength.
    (D) They are formed from oscillating electric and magnetic fields.
    (E) All of the statements are true.

53. What is true of all combinations of resistors arranged in parallel?

    I. Current splits down each leg of the combination.

    II. Voltage is constant across each leg of the combination.

    III. Resistance is the same for each leg of the combination.

    (A) I and II
    (B) I and III
    (C) II and III
    (D) II only
    (E) I, II, and III

54. How much current flows when 3.6 µC of charge flows in 10 ms?

    (A) 0.36 mA
    (B) 0.36 A
    (C) 3.6 A
    (D) 36 A
    (E) 360 A

GO ON TO THE NEXT PAGE. ➡

Questions 55-58 are based on the circuit below.

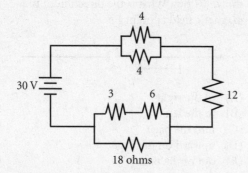

55. What current flows through the 12 Ω resistor?

    (A) 2.5 A
    (B) 0.4 A
    (C) 1.5 A
    (D) 0.75 A
    (E) 1.25 A

56. Find the potential difference across the 18 Ω resistor.

    (A) 30 V
    (B) 27 V
    (C) 18 V
    (D) 15 V
    (E) 9 V

57. How much total power is dissipated by the two 4 Ω resistors?

    (A) 24 W
    (B) 2.0 W
    (C) 4.5 W
    (D) 12 W
    (E) 0.04 W

58. Where should you add a 12 Ω resistor to produce the largest increase in total current in this circuit?

    (A) in parallel with one of the 4 Ω resistors
    (B) in parallel with the 12 Ω resistor
    (C) in series with the 12 Ω resistor
    (D) in parallel with the 18 Ω resistor
    (E) Adding a resistor will only decrease the current.

59. All of the following statements about current are true EXCEPT:

    (A) The direction of current is defined as the direction of the positive charge flow.
    (B) The more charge that flows per second, the larger the current.
    (C) Current is produced when electrons in metals flow from negative to positive.
    (D) The units for current are Coulombs·seconds.
    (E) Current cannot flow around an "open loop."

60. Which of the following is equivalent to a volt (v)?

    (A) J/C
    (B) J·s/C
    (C) V·J
    (D) V/s
    (E) J/s

61. A 10-ohm resistor and a 20-ohm resistor are connected in parallel with each other to a battery. Which of the following is true?

    (A) The current flowing through the resistors is the same.
    (B) The resistor values must be equal for the circuit to work.
    (C) The potential difference across the two resistors is the same.
    (D) The power used by the two resistors is the same.
    (E) More current will flow through the 20-ohm resistor.

62. The capacitance of a parallel-plate capacitor may be increased by:

    (A) increasing the distance between the plates
    (B) increasing the voltage
    (C) increasing the conductivity of the material between the plates
    (D) increasing the area of the plates
    (E) decreasing the voltage

63. Two charges of +Q are 1 cm apart and exert forces on each other. If one of the charges is replaced by a charge of –Q, the magnitude of the force between the two charges now is:

   (A) zero
   (B) smaller
   (C) larger
   (D) the same
   (E) changed only if the magnitudes of the charges are equal to each other

64. A negative charge moving to the right enters a uniform magnetic field that is directed into the page, as shown below. What will be the initial change in direction of the motion of the charge as it enters the field?

   (A) upward on the page
   (B) downward on the page
   (C) into the page
   (D) out of the page
   (E) decelerating to the right

65. You have available a power supply with potential difference, V, and several identical capacitors. Which arrangement of capacitors will store the greatest amount of charge when charged by the power supply?

   (A) one single capacitor
   (B) two of the capacitors in series
   (C) three of the capacitors in series
   (D) two of the capacitors in parallel
   (E) three of the capacitors in parallel

66. A long wire carries a current to the right, as shown below. What is the direction of the magnetic field at point P?

   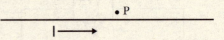

   (A) to the right
   (B) to the left
   (C) into the page
   (D) upward on the page
   (E) out of the page

67. Which of the following is an electromagnetic frequency visible to humans?

   (A) 5×10^4 Hz
   (B) 5×10^8 Hz
   (C) 5×10^{14} Hz
   (D) 5×10^{20} Hz
   (E) 5×10^{30} Hz

68. Two capacitors, with values 3 µF and 6 µF, are connected in parallel and charged with a 9-volt battery. How much total charge is stored in the two capacitors?

   (A) 1 µC
   (B) 4.5 µC
   (C) 11 µC
   (D) 18 µC
   (E) 81 µC

69. Which of the following does not obey an inverse square law, i.e., strength or intensity decreases with the inverse square of distance?

   (A) light intensity
   (B) sound intensity
   (C) gravitational field
   (D) electric field
   (E) electric potential

70. Which of the following quantities does not normally increase with an increase in temperature?

   (A) electrical resistance
   (B) speed of sound in air
   (C) density of a gas
   (D) pressure of a gas at constant volume
   (E) linear dimensions of solid objects

END OF SECTION I

AP Physics
PRACTICE TEST II

PRACTICE TEST II
AP PHYSICS B

SECTION II

Time—90 minutes
Questions—6
Points—90

Directions: This portion of the test consists of six free-response questions and should be taken in a 90-minute time period. The point value of each question is indicated. Not all parts of each question have equal weight. A calculator and list of formulas are allowed.

1. (15 points)

   Two children are playing on a pulley system. This pulley system allows one child to jump off a platform while pulling the other child across the horizontal platform.

   (a) If the jumping child has a mass of 65 kg and the sliding child's mass is 85 kg, what maximum value for the coefficient of static friction will allow the larger child to begin sliding? (4 points)

   (b) Once sliding, the coefficient of kinetic friction becomes 0.20. What will the children's acceleration be? (4 points)

   (c) How much additional mass should the sliding child hold in order to maintain a constant speed? (4 points)

   (d) After the children play with the pulley for a while, their father comes out and offers to pull both of them by standing on the ground and pulling on the rope. Using the same coefficient of static friction as in part (a), determine the force the father must exert to start the children sliding. (3 points)

GO ON TO THE NEXT PAGE. ➡

KAPLAN 317

AP Physics
PRACTICE TEST II

2. A pellet gun fires a bullet into a stationary block of wood that is attached to a spring on a frictionless surface. When the bullet enters the wood, it remains inside, and the bullet and block enter into simple harmonic motion with amplitude = 11.0 cm. The bullet (m= 5 g) was initially traveling at 650 m/s before hitting the block of wood (m = 2.5 kg).

(a) What is the spring constant of the spring? (6 points)

(b) What is the total energy of the system after the collision? (2 points)

(c) What is the maximum acceleration of the bullet/block system once it begins its oscillation? (4 points)

(d) Where will the bullet and block reach 0 velocity? (3 points)

318 **KAPLAN**

AP Physics
PRACTICE TEST II

3. (15 points)

A car travels around a circular turn of radius 50 meters while maintaining a constant speed of 15 m/s.

(a) Draw and clearly label a force diagram of all forces acting on the car in this problem, using the dot below to represent the car. (4 points).

•

(b) Find the minimum value for the coefficient of friction necessary to keep the car on the road. (4 points)

(c) In icy conditions, the coefficient of friction drops to 0.15. Now find the maximum safe speed for making the turn. (3 points)

(d) If the flat roadway were replaced by a banked one, what angle of incline would be necessary for the car to make the turn at this same speed without requiring the use of friction? (4 points)

GO ON TO THE NEXT PAGE. ➡

KAPLAN 319

AP Physics
PRACTICE TEST II

4. (15 points)

Use the circuit below to answer parts 4a through 4d.

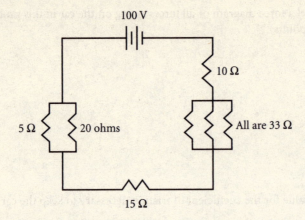

(a) Find the total equivalent resistance. (5 points)

(b) Find the total current in the circuit. (4 points)

(c) Find the voltage drop across the 10 Ω resistor. (3 points)

(d) Through which resistor does the least current flow, and what is the value of that current? (3 points)

5. (15 points)

A 1.0 m piece of wire is coiled into 200 loops and attached to a voltage source as shown.

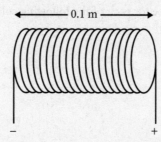

(a) Find the strength of the magnetic field inside the coil if V = 100 V and R = 40 Ω. (5 points)

(b) Which direction does the magnetic field point? (3 points)

(c) The wire is then uncoiled and rewrapped so that the cross-sectional area of the coil is twice what it was previously, though the length stays the same. What is the new magnetic field strength inside the coil? (4 points)

(d) The entire coil of wire is now placed into an external magnetic field as shown below. Which direction will the coil first begin to rotate? (3 points)

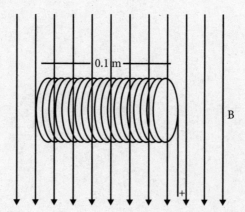

AP Physics
PRACTICE TEST II

6. (15 points)

White light is shone perpendicularly onto a film of unknown thickness, with index of refraction n = 1.5.

(a) How thick would the film have to be for each of the following colors of light to be absorbed due to destructive interference? (10 points)

Red (f = 4×10^{14} Hz)

Green (f = 5.45×10^{14} Hz)

Blue (f = 6.38×10^{14} Hz)

(b) Draw a ray diagram that indicates the path of the red light from part (a) into and out of the film and justify the fact that red will not be seen in the reflected light. (5 points)

white light ↓   air

film

air

PRACTICE TEST II
ANSWER KEY

Section I—Multiple-Choice

1. A	19. A	37. D	55. C
2. C	20. D	38. B	56. E
3. A	21. C	39. D	57. C
4. B	22. B	40. D	58. B
5. C	23. A	41. D	59. D
6. C	24. D	42. A	60. A
7. B	25. B	43. A	61. C
8. C	26. B	44. D	62. D
9. C	27. D	45. A	63. D
10. C	28. E	46. E	64. B
11. D	29. B	47. D	65. E
12. A	30. B	48. D	66. E
13. D	31. D	49. A	67. C
14. C	32. B	50. D	68. E
15. B	33. B	51. B	69. E
16. D	34. C	52. E	70. C
17. A	35. C	53. A	
18. B	36. E	54. A	

Answers and Explanations to Practice Test II

SECTION I: MULTIPLE-CHOICE

Note: 10 m/s^2 is used for g in most cases.

1. **(A)** A horizontal line on a displacement-time graph would have a zero slope. Since the slope of displacement vs time is velocity, the velocity is zero.

2. **(C)** Since they fall from the same height, they will reach the ground after the same amount of time and will have the same vertical velocity. However, the ball that is thrown horizontally has the additional horizontal component of velocity when it lands, giving ball A the greater speed upon impact.

3. **(A)** Since the wheel drops, its initial vertical velocity is zero. Examining only the wheel's vertical motion:

$$s_y = v_{oy}t + \frac{1}{2}at^2$$

$$-490 \text{ m} = \left(\frac{1}{2}\right)(-10 \text{ m/s}^2)t^2$$

Rounding the 490 to 500, t^2 equals approximately 100, and t = 10 seconds.

KAPLAN 325

AP Physics
ANSWERS AND EXPLANATIONS TO PRACTICE SET II

4. **(B)** Since the stone is dropped, its initial velocity is zero.

$$s_y = v_{oy}t + \frac{1}{2}at^2$$

$$-30 \text{ m} = \left(\frac{1}{2}\right)(-10 \text{ m/s}^2)t^2$$

$$t^2 = 6$$

Now, use $v_f = v_o + at$, and $v_f = (-10 \text{ m/s}^2)t$

The nearest estimate for this is between 20 and 30 m/s, so we choose answer **(B)**. (Note: It would be faster to use the equation $v_f^2 = v_o^2 + 2as$.)

5. **(C)** Set the applied force equal to the retarding friction force:

$$F_{app} = F_{fr} = \mu F_{normal}$$

On a level surface, the normal force equals the weight, so:

$$F_{app} = \mu mg$$

$$\mu = 40 \text{ N}/(10 \text{ kg})(10 \text{ m/s}^2) = 0.4.$$

6. **(C)** An object traveling in a circle may have constant speed, but its velocity is constantly changing as the object constantly changes direction. A constant centripetal force causes this change in direction to maintain circular motion.

7. **(B)** From the formula for centripetal force:

$$F_c = mv^2/R$$

When the radius is doubled, the force must be one half as much to maintain the same speed.

8. **(C)** By Newton's Third Law of Motion, the two objects exert forces on each other that are equal in magnitude but opposite in direction.

9. **(C)** The centripetal force to keep the car in circular motion in the turn is provided by the friction force alone, assuming a level road.

$$F_c = F_{fr}$$

$$mv^2/R = \mu mg$$

$$R = v^2/\mu g = (10 \text{ m/s})^2/(0.5)(10 \text{ m/s}^2) = 20 \text{ m}.$$

326 KAPLAN

AP Physics

ANSWERS AND EXPLANATIONS TO PRACTICE SET II

10. **(C)** In segment C, the velocity is zero.

11. **(D)** The slope of segment D is larger in magnitude than any other labeled segment, and the slope of a velocity vs time graph is acceleration.

12. **(A)** In segment A, the value is non-zero, so the object is moving, but the slope is zero, so it is not accelerating.

13. **(D)** Since the question asked for speed, not velocity, both positive and negative values need to be considered. The object has a speed of 4 m/s at times of 3 and 10 seconds.

14. **(C)** The area under the velocity vs time graph represents change in position, or distance traveled. Considering the shape in two segments—a rectangle with area 15 meters and a triangle with area 4.5 meters—gives a total distance of 19.5 m.

15. **(B)** First, consider the gravitational force that provides the centripetal force to keep each satellite in orbit:

$$F_G = F_c$$
$$GMm/R^2 = mv^2/R$$

The velocity in orbit equals: $\left(\sqrt{\dfrac{GM}{R}} \right)$, where M is the mass of Earth.

Thus, a satellite orbiting farther from Earth has a smaller magnitude of velocity.

Since centripetal acceleration equals v^2/R, by this relationship the centripetal acceleration would be larger for an object orbiting nearer to Earth.

16. **(D)** The effective value of g will increase in an elevator accelerating upward or in the Space Shuttle accelerating upward, thus increasing the weight. The object in free fall would seem weightless.

17. **(A)** Only diagram A is a correct free body diagram. Note that in D both the weight force and its components are shown as forces, which is redundant and effectively doubles the weight.

18. **(B)** From the form $x(t) = A \cos\omega t$, the amplitude, A, would be 0.50 meters in this case.

KAPLAN 327

AP Physics
ANSWERS AND EXPLANATIONS TO PRACTICE SET II

19. **(A)** Since $\omega = 2\pi f$ and $f = 1/T$, from the equation $x(t) = A\cos\omega t$:

$$\omega = 2\pi/T = 10$$

$$T = 2\pi/10 \text{ seconds.}$$

20. **(D)** From the equation, $A = 0.5$ m.

Maximum potential energy of a spring $= \dfrac{1}{2}kA^2 = \left(\dfrac{1}{2}\right)(100 \text{ N/m})(0.5 \text{ m})^2 = 12.5$ J.

21. **(C)** Maximum potential energy of the spring is converted to maximum kinetic energy when the mass moves through the equilibrium position:

$$\frac{1}{2}kA^2 = \frac{1}{2}mv^2$$

$$v = (kA^2/m)^{\frac{1}{2}} = [(100 \text{ N/m})(0.5 \text{ m})^2/(1 \text{ kg})]^{\frac{1}{2}} = 5 \text{ m/s}$$

This result can also be obtained by using the relationship $v = \omega R$, which in this case would be $v = (10 \text{ rad/s})(0.5 \text{ m}) = 5$ m/s.

22. **(B)** The area under a "force vs time" graph would represent change in momentum, or impulse.

23. **(A)** The direct increase of position and time on the graph would represent motion with a constant velocity (i.e., slope is constant).

24. **(D)** The linear relationship of the graph could represent velocity of a falling object, if the initial velocity was zero: $v = at$.

25. **(B)** Since the energy of light is directly proportional to frequency of light by the equation $E = hf$, the slope of the line would be Planck's constant, h.

26. **(B)** Ranked in order from smallest wavelength to largest:

gamma, x-rays, visible light, radio waves.

27. **(D)** Critical angle only occurs for light moving from a medium of higher index of refraction into a medium of lower index. Diamond has a higher index of refraction than air.

ANSWERS AND EXPLANATIONS TO PRACTICE SET II

AP Physics

28. **(E)** Examine the equation for diffraction: $m\lambda = d\sin\theta$

    Regardless of the slit width in the grating, the angle through which the color is diffracted to produce the first order maximum will be largest when the wavelength is largest. Of the colors listed, red has the longest wavelength.

29. **(B)** Only choice (B) is concave.

30. **(B)** The bubble film has varying thickness as gravity pulls the soap downward. Reflections of light from the outer surface of the bubble interfere with reflections from within the film itself, causing a variation of visible colors.

31. **(D)** Diffraction occurs as waves pass through an opening that has a width of the same order of magnitude as the wavelength of the waves. Light waves have wavelengths much too small for the light to be diffracted by an open doorway. The smaller frequency of sound waves (and smaller velocity) translates to much larger wavelengths, which might be diffracted.

32. **(B)** Use the equation for diffraction: $m\lambda = d\sin\theta$.

    You can see from the equation that as wavelength increases, the angle through which the bright lines are spread also increases.

33. **(B)** Using the thin lens equation:

    $$1/f = 1/d_o + 1/d_i$$
    $$\frac{1}{20} = \frac{1}{15} + 1/d_i, \text{ or } 1/d_i = \frac{1}{20} - \frac{1}{15}$$

    Using the handy "product over sum" rule for this expression:

    $$d_i = -(20)(15)/(20-15) = -60 \text{ cm}$$

    Magnification,

    $$M = -h_i/h_o = d_i/d_o$$
    $$h_i = h_o d_i/d_o = (10)(-60)/15 = 40 \text{ cm}.$$

KAPLAN 329

AP Physics
ANSWERS AND EXPLANATIONS TO PRACTICE SET II

34. **(C)** First with the initial conditions, the magnification is $\frac{1}{2}$, so image distance is one-half the object distance:

$$M_1 = h_i/h_o = \frac{1}{2} = -d_i/d_o$$

Therefore: $d_i = -Md_o$

Now substituting into the thin lens equation when $M = \frac{1}{4}$:

$$1/-f = 1/d_o + 1/d_i = 1/d_o + 1/-\left(\frac{1}{4}\right)(d_o)$$

$$-1/f = 1/d_o - 4/d_o = -3/d_o$$

$$d_o = 3f.$$

35. **(C)** The wavelength of light in the film, in combination with the thickness of the film, will determine the phase of the reflected light as it returns through the film to interfere with reflected light from the top surface.

36. **(E)** A real image only can be formed after light rays pass through the lens or reflect from a mirror in such a way that they converge to form an image. Light passing through a concave lens will always diverge, so a real image will not be formed.

37. **(D)** Total pressure equals atmospheric pressure plus water pressure: $P_t = P_a + \rho gh$. Using 1,000 kg/m^3 as the density of fresh water, the water pressure is approximately one atmosphere for each 10 meters of depth:

$$P_{water} = (1{,}000 \text{ kg/m}^3)(10 \text{ m/s}^2)(10 \text{ m}) = 10^5 \text{ Pa}$$

Thus, 3 atmospheres would include one atmosphere of air plus 20 meters of water.

It's helpful to remember that each 10 m of water depth is equivalent to about 1 atm of pressure.

38. **(B)** If the buoyant force were as great as the weight of the wood, the wood would float. If the wood sinks, it is not able to displace enough water such that the water weighs as much as the wood, so the buoyant force is less than the wood's weight.

39. **(D)** In a transverse wave, the individual particles of the medium move perpendicular to the direction of travel of the wave itself.

330 **KAPLAN**

ANSWERS AND EXPLANATIONS TO PRACTICE SET II

AP Physics

40. **(D)** The rate at which heat is conducted through a medium does not depend upon the specific heat of the material:

$$\frac{Q}{t} = \frac{kA\Delta T}{d}, \text{ where } k = \text{conductivity, } A = \text{area, } d = \text{thickness, and } \Delta T \text{ is}$$

temperature gradient.

41. **(D)** Sound waves are not electromagnetic.

42. **(A)** As a sound approaches a stationary observer, the sound will appear to the observer to have higher pitch or higher frequency than actual.

43. **(A)** Bending of waves around a barrier is called diffraction.

44. **(D)** Hearing 4 beats per second only indicates a difference of 4 Hz in the two frequencies. Without other information, one can only determine that the unknown frequency is either 252 or 260 Hz.

45. **(A)** An adiabatic system is one is which $Q = 0$, or no heat enters or leaves the system.

46. **(E)** Bernoulli's equation is based on conservation of energy. To see this, take the equation and multiply through by volume:

$$P + \rho g h + \frac{1}{2} \rho v^2 = \text{constant}$$

$$PV + \rho g h V + \frac{1}{2} \rho V v^2 = \text{constant}(V)$$

$$\text{work} + mgh + \frac{1}{2} mv^2 = \text{constant.}$$

47. **(D)** Use the fluid continuity equation, $A_1 v_1 = A_2 v_2$. If the second diameter is half the original, then the radius is also half the original. Cutting the radius in half will cut the area to one-fourth. Thus, if the second area is one-fourth the original, then the velocity must become four times as great.

48. **(D)** As light moves from one medium to another, its frequency remains the same. Moving from higher index of refraction to lower index of refraction, the velocity and wavelength of the light will both increase.

KAPLAN 331

AP Physics
ANSWERS AND EXPLANATIONS TO PRACTICE SET II

49. **(A)** If the temperature of an ideal gas increases at constant volume, then the pressure also increases, since Kelvin temperature is directly proportional to pressure. If the volume does not change, no work is done. Since temperature is a measure of average translational kinetic energy of molecules, they both increase together.

50. **(D)** Total fluid pressure depends on atmospheric pressure, fluid density, and depth:

$$P_{total} = P_{atm} + \rho gh$$

Surface area is not a factor.

51. **(B)** Since the decibel level is on a logarithmic scale, an increase of 10 on the scale indicates an increase in intensity that is ten times.

52. **(E)** All statements are true.

53. **(A)** For resistors arranged in parallel, current will split according to the inverse of the amount of resistance in each branch, but the potential difference across each branch will be the same. However, resistance will not necessarily be the same in each branch.

54. **(A)** $I = Q/t$

$I = (3.6 \times 10^{-6} \text{ C})/(10 \times 10^{-3} \text{ s}) = 0.36 \times 10^{-3} \text{ A or } 0.36 \text{ mA.}$

55. **(C)** First, find the equivalent resistance for the circuit.

Beginning with the parallel branch with 3 resistors, the 3 ohms and 6 ohms in series add to 9 ohms. This 9 ohms in parallel with 18 ohms is equivalent to:

$(9)(18)/(9 + 18) = 6$ ohms

(Note: Since there are just two parallel branches, the "product over sum" rule will work here.)

Next, the two 4 ohm resistors in parallel are equivalent to:

$(4)(4)/(4 + 4) = 2$ ohms

Now we have 12, 2, and 6 in series for a total of 20 ohms. Using Ohm's Law:

$V = IR$, for the circuit, $I = V/R = 30/20 = 1.5$ amps

All of this current flows through the 12 ohm resistor.

332   **KAPLAN**

AP Physics
ANSWERS AND EXPLANATIONS TO PRACTICE SET II

56. **(E)** The current splits in these parallel branches so that twice as much current flows through the 9 ohm branch (with the 3 and 6 ohm resistors). Thus, the 1.5 amp current will split so that 1 A flows through the 3 and 6 ohm resistors and 0.5 A. flows through the 18 ohm resistor. Using Ohm's Law,

$$V = IR = (0.5 \text{ A.})(18 \ \Omega) = 9 \text{ volts.}$$

57. **(C)** We can do the power for each resistor separately and then add power. Each resistor receives half the total current, or 0.75 A. Using $P = I^2R = (0.75 \text{ A})^2(4 \ \Omega)$, the power used by each resistor is 2.25 watts, for a total of 4.5 watts for both resistors. We could alternately use the total resistance for the two, which is 2 ohms, in the same equation with the total current, which is 0.75 amps, to get the same result.

58. **(B)** To increase the total current in the circuit, we want to create the smallest total resistance possible. Putting the 12 ohm resistor in parallel with the existing 12 ohm resistor will accomplish this, reducing the equivalent resistance of that part of the circuit to 6 ohms.

59. **(D)** Current is the rate of flow of charge, measured in Coulombs per second, or Amps.

60. **(A)** A volt is equivalent to a Joule/Coulomb, since voltage is potential energy per unit charge.

61. **(C)** Two resistors connected in parallel must have the same potential difference across them, due to conservation of energy in the circuit, or Kirchoff's Loop Rule.

62. **(D)** The capacitance of a parallel plate is increased by increasing plate area:

$$C = \frac{kA}{d}$$

63. **(D)** In the equation, changing the sign of one of the charges will not change the magnitude of the force, only the direction: $F = kQ_1Q_2/R^2$.

64. **(B)** Using the right hand rule, or the vector cross product of velocity and field, the force would be downward on the page, so the particle will begin to move downward as it moves to the right.

KAPLAN 333

AP Physics
ANSWERS AND EXPLANATIONS TO PRACTICE SET II

65. **(E)** The amount of charge stored is directly proportional to the capacitance, so we want to have the greatest total capacitance. The total capacitance for capacitors connected in parallel is found by adding their values, so connecting three capacitors in parallel would give the greatest value.

66. **(E)** Using the right hand rule, the field would curve around the wire, outward from the page at P.

67. **(C)** Using the speed of light, 3×10^8 m/s, and the knowledge that visible light ranges from about 400 to 750 nm, solve for an approximate value for frequency using $c = f\lambda$.

$$f = (3 \times 10^8 \text{ m/s })/(500 \times 10^{-9} \text{ m}) = 0.6 \times 10^{15} \text{ or } 6 \times 10^{14} \text{ Hz}$$

Answer **(C)** is closest.

68. **(E)** First, determine the equivalent capacitance for the two capacitors:

$$C_{eq} = C_1 + C_2 = 9 \text{ μF}$$

Then, find the charge from the equation $Q = CV$.

$$Q = (9 \times 10^{-6} \text{ F})(9 \text{ v}) = 81 \times 10^{-6} \text{ or } 81 \text{ μC}.$$

69. **(E)** All the quantities listed obey an inverse square law, i.e., the quantities decrease with the inverse square of the distance, except electric potential:

$$V = kQ/R.$$

70. **(C)** An increase in temperature should decrease the density of an ideal gas, while an increase in temperature should increase each of the other quantities.

334 **KAPLAN**

AP Physics

ANSWERS AND EXPLANATIONS TO PRACTICE SET II

SECTION II: FREE RESPONSE

1. (15 points)

    Two children are playing on a pulley system. This pulley system allows one child to jump off a platform while pulling the other child across the horizontal platform.

    (a) If the jumping child has a mass of 65 kg and the sliding child's mass is 85 kg, what maximum value for the coefficient of static friction will allow the larger child to begin sliding? (4 points)

    Answer:

    Assuming no acceleration, the weight of the 65 kg child should exert enough force to exactly balance the friction force between the sliding child and platform.

    $$\Sigma F = 0 = mg - F_{fr} = mg - \mu F_N$$

    $$(65 \text{ kg})(9.8 \text{ m/s}^2) - \mu(85 \text{ kg})(9.8 \text{ m/s}^2) = 0$$

    $$\mu = 0.76.$$

    (b) Once sliding, the coefficient of kinetic friction becomes 0.20. What will the children's acceleration be? (4 points)

    Answer:

    Now, the net force is not zero but produces an acceleration.

    $\Sigma F = ma$, where $(M + m)$ is the total mass of the system, or sum of both children's masses.

    $$mg - \mu Mg = (m + M)a$$

    $$(65 \text{ kg})(9.8 \text{ m/s}^2) - (0.20)(85 \text{ kg})(9.8 \text{ m/s}^2) = (150 \text{ kg})(a).$$

    $$a = 3.14 \text{ m/s}^2 \cdot$$

    (c) How much additional mass should the sliding child hold in order to maintain a constant speed? (4 points).

    Answer:

    We'll add enough mass to the sliding child's mass, using the same coefficient of friction, to make the net force equal zero.

    $$\Sigma F = 0 = (65 \text{ kg})(9.8 \text{ m/s}^2) - (0.20)(85 \text{ kg} + M)(9.8 \text{ m/s}^2)$$

    $$M = 240 \text{ kg}$$

KAPLAN 335

AP Physics
ANSWERS AND EXPLANATIONS TO PRACTICE SET II

(d) After the children play with the pulley for a while, their father comes out and offers to pull both of them by standing on the ground and pulling on the rope. Using the same coefficient of static friction as in part (a), determine the force the father must exert to start the children sliding. (3 points)

Answer:

Again, we'll use 0.76, the coefficient of static friction we found above, and the total mass of the two children.

$$\Sigma F = 0 = F - F_{fr} = F - \mu F_N$$

$$F - (0.76)(150 \text{ kg})(9.8 \text{ m/s}^2) = 0$$

$$F = 1{,}117 \text{ N}$$

2. (15 points)

A pellet gun fires a bullet into a stationary block of wood that is attached to a spring on a frictionless surface. When the bullet enters the wood, it remains inside, and the bullet and block enter into simple harmonic motion with amplitude = 11.0 cm. The bullet (m = 5 g) was initially traveling at 650 m/s before hitting the block of wood (m = 2.5 kg).

(a) What is the spring constant of the spring? (6 points)

Answer:

To answer this, we need the fact that the kinetic energy of the bullet/block system is converted into spring potential energy like this:

$$KE = \frac{1}{2} mv^2 = PE_{spring} = \frac{1}{2} kx^2$$

We know the mass of the bullet/block (b/b) system: 2.505 kg (mass of the block plus the mass of the bullet). So the only thing we're missing is the speed of the bullet/block system. To find this, we use the Law of Conservation of Momentum. In this case:

$$P_i = m_{bullet} \cdot V_{bullet} = (0.005 \text{ kg})(650 \text{ m/s}) = 3.25 \text{ kg·m/s} = P_f = m_{b/b} \cdot v_{b/b} = (2.505 \text{ kg})(V_{b/b})$$

Solving for the speed, we get:

$$v_{b/b} = \frac{3.25}{2.505} = 1.30 \text{ m/s}$$

Now we return to the Conservation of Energy:

$$\frac{1}{2}(2.505 \text{kg})(1.30 \text{ m/s})^2 = 2.12\text{J} = \frac{1}{2}k(0.11\text{m})^2$$

So, the spring constant (k) = 350.4 N/m.

(b) What is the total energy of the system after the collision? (2 points)

Answer:

We already found this in part (a). The total energy of the system = 2.12 J

(c) What is the maximum acceleration of the bullet/block system once it begins its oscillation? (4 points)

Answer:

Maximum acceleration occurs when a maximum force is present. This happens when the spring is either fully stretched or fully compressed. Using F=(–)kx, we find the maximum force = (350.4 N/m)(0.11 m) = 38.5 N toward equilibrium

This force, acting on a mass of 2.505 kg produces an acceleration of

$$a = \frac{F}{m} = \frac{38.5 N}{2.505 kg} = 15.4 \text{ m/s}^2 \text{ toward equilibrium}$$

(d) Where will the bullet and block reach 0 velocity? (3 points)

Answer:

The b/b system comes to rest at either end of its oscillation, that is, at full compression and at full extension.

3. (15 points)

A car travels around a circular turn of radius 50 meters while maintaining a constant speed of 15 m/s.

(a) Draw and clearly label a force diagram showing all forces acting on the car in this problem. (4 points)

Answer:

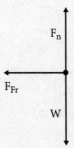

AP Physics
ANSWERS AND EXPLANATIONS TO PRACTICE SET II

(b) Find the minimum value for the coefficient of friction necessary to keep the car on the road. (4 points)

Answer:

Setting friction force equal to centripetal force $\mu mg = \dfrac{mv^2}{r}$ leads to an expression for the coefficient of friction: $\mu = \dfrac{v^2}{gr} = \dfrac{(15 \text{ m/s})^2}{(9.8 \text{ m/s}^2)(50 \text{ m})} = 0.46$

(c) In icy conditions, the coefficient of friction drops to 0.15. Now find the maximum safe speed for making the turn. (3 points)

Answer:

Again, using the relationship between speed and coefficient of friction:

$$v = \sqrt{\mu gr} = \sqrt{(0.15)(9.8 \text{ m/s}^2)(50 \text{ m})} = 8.6 \text{ m/s}$$

(d) If the flat roadway were replaced by a banked one, what angle of incline would be necessary for the car to make the turn at this same speed, without requiring the use of friction? (4 points)

Answer:

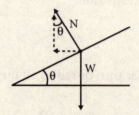

When we use a banked curve instead of friction, a component of the normal force now provides the centripetal force:

$N\sin\theta = mv^2/r$

$N\cos\theta = mg$

Combining these:

$$\tan\theta = \dfrac{v^2}{gr} = \dfrac{(8.6 \text{ m/s})^2}{(9.8 \text{ m/s}^2)(50 \text{ m})}$$

Solving:

$\tan\theta = 0.15$

$\theta = 8.53°$

4. (15 points)

Use the circuit below to answer parts 4a through 4d.

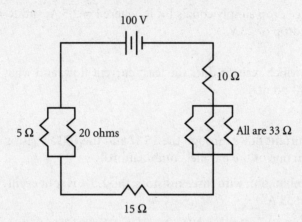

(a) Find the total equivalent resistance. (5 points)

Answer:

In the 3-resistor parallel branch:

$1/R_3 = \dfrac{1}{33} + \dfrac{1}{33} + \dfrac{1}{33}$

$R_3 = 11$ ohms

In the 2-resistor parallel branch:

$1/R_2 = \dfrac{1}{5} + \dfrac{1}{20}$

$R_2 = 4$ ohms

Now add up the two single resistors with the two branches:

$R_{eq} = 10 + 15 + 11 + 4 = 40$ ohms

(b) Find the total current in the circuit. (4 points)

Answer:

Once we know the total resistance (40 Ω) and the total voltage (100 V), we apply Ohm's Law:

$V = IR$

$100\ V = I\ (40\ \Omega)$

$I = 2.5$ A.

AP Physics
ANSWERS AND EXPLANATIONS TO PRACTICE SET II

(c) Find the voltage drop across the 10 Ω resistor. (3 points)

Answer:

The voltage drop simply equals I × R, where I = 2.5 A, and R = 10 Ω. This gives a potential drop of 25 V.

(d) Through which resistor does the least current flow, and what is the value of that current? (3 points)

Answer:

A 2.5 A current flows through the 15 Ω and the 10 Ω resistor each, so the answer must lie in one of the parallel combinations.

In the combination with three resistors, the 2.5 A is split evenly where each leg gets 2.5/3 = 0.833 A.

In the other parallel combination, we have an equivalent resistance of 4 Ω through which 2.5 A flow. Thus, the potential drop across the combination is 10 V (V = IR). A voltage drop of 10 V across 5 Ω produces a current of 2 A.

A voltage drop of 10 V across 20 Ω produces a current of 0.5 A. This accounts for the necessary 2.5 A in the circuit and tells us that the resistor with the least current is the 20 Ω resistor.

5. (15 points)

A 1.0 m piece of wire is coiled into 200 loops and attached to a voltage source as shown.

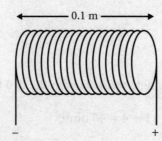

(a) Find the strength of the magnetic field inside the coil if V = 100 V and R = 40 Ω. (5 points)

Answer:

Given the values of V and R, the current, I, flowing in the wire is V/R = 100 V/40 Ω = 2.5 A. There are 200 turns in a length of 0.1 meters, so the coil density (turns/length) = 2,000 m^{-1}.

Thus, the magnetic field inside the solenoid is

$$B = \mu_0 n I = (4\pi \cdot 10^{-7} \text{ T} \cdot \text{m/A})(2000 \text{ m}^{-1})(2.5 \text{ A}) = 0.00628 \text{ T}.$$

ANSWERS AND EXPLANATIONS TO PRACTICE SET II

(b) Which direction does the magnetic field point? (3 points)

Answer:

According to the right-hand rule, the magnetic field inside the solenoid (which makes it act like a bar magnet) will point to the right in the drawing.

(c) The wire is then uncoiled and rewrapped so that the cross-sectional area of the coil is twice what it was previously, though the length stays the same. What is the new magnetic field strength inside the coil? (4 points)

Answer:

The only value that changes is the number of turns per meter. Originally, there were 200 turns that used up 1 meter (100 cm) of wire. Thus, each turn consisted of 0.5 cm of wire. This is the circumference, which is proportional to the radius. The area is proportional to the square of the radius. If we double the area, we need to increase the radius by a factor of $\sqrt{2}$, and with it also the circumference by the same factor. Since now each turn becomes longer by $\sqrt{2}$, the number of turns that we can get out of 1 m of wire is decreased by $\sqrt{2}$. Since everything else remains the same as in part A, and only the number of turns decreases by $\sqrt{2}$, the magnetic field will also decrease by $\sqrt{2}$. So, $B = \dfrac{0.00628 T}{\sqrt{2}} = 0.00444$ T.

(d) The entire coil of wire is now placed into an external magnetic field as shown below. Which direction will the coil first begin to rotate? (3 points)

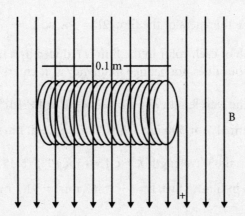

AP Physics
ANSWERS AND EXPLANATIONS TO PRACTICE SET II

Answer:

Only the component of the current that flows perpendicular to the external field will cause any force to be produced. Let's look at the top and bottom of the coil. Across the top, the right-hand rule tells us that there will be a force produced to the right of the coil. Across the bottom, there will be a force produced to the left of the coil, which will cause the coil to begin rotating in a clockwise direction.

6. (15 points)

White light is shone perpendicularly onto a film of unknown thickness, with index of refraction n = 1.5.

(a) How thick would the film have to be for each of the following colors of light to be visible due to constructive interference? (10 points)

Red (f = 4×10^{14} Hz)

Green (f = 5.45×10^{14} Hz)

Blue (f = 6.38×10^{14} Hz)

Answer:

Constructive interference occurs when waves return from the reflected surfaces (top and bottom) in phase with each other. The reflected rays from the top surface are switched out of phase upon reflection due to the higher index of refraction.

Thus, the path of the light passing through the film and back must be $\frac{1}{2}$ of a wavelength longer than the path of the light that reflects from the surface. Of course, some other path lengths would also produce destructive interference, but this is the shortest and describes the smallest thickness that produces the desired effect.

Thus, if L is the thickness of the film, $2L = \frac{1}{2}\lambda$, so $L = \frac{1}{4}\lambda$.

The wavelength of each color in the film is reduced by a factor of n, since the light reduces speed but does not change frequency as it enters the film.

For red light, the wavelength $\lambda = c/f = (3 \times 10^8)/(4 \times 10^{14}) = 750$ nm. Wavelength in film is 750 nm/1.5 or 500 nm. Thus, $L = \left(\frac{1}{4}\right)(500 \text{ nm}) = 125$ nm.

For green light, the wavelength $\lambda = c/f = (3 \times 10^8)/(5.45 \times 10^{14}) = 550$ nm in air or 550 nm/1.5 in film. Thus, $L = \left(\frac{1}{4}\right)(367 \text{ nm}) = 91.7$ nm.

For blue light, the wavelength $\lambda = c/f = (3 \times 10^8)/(6.38 \times 10^{14}) = 470$ nm in air, or 313.3 nm in film. Thus, $L = \left(\frac{1}{4}\right)(313.3 \text{ nm}) = 78.3$ nm.

342 **KAPLAN**

(b) Draw a ray diagram that indicates the path of the red light from part (a) into and out of the film and justify the fact that red will be seen in the reflected light. (5 points)

Answer

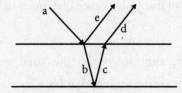

Note: The rays drawn are angled from the actual perpendicular to make it easier to examine individual changes. As the incident ray [a] reflects from the surface, it changes phase, so that rays a and e are out of phase with each other. Rays b, c, and d remain in phase with the original incident ray, since neither refraction nor the reflection from the lower surface will cause a phase change. Thus, to be able to see the red light from the film, the ray emerging from the film needs to be switched out of phase also, which can only occur if the path length through the film is one-half wavelength, as calculated above.

AP Physics

ANSWERS AND EXPLANATIONS TO PRACTICE SET II

COMPUTING YOUR PRACTICE TEST SCORE

To approximate your score for each test, you will need to equalize the two sections for a total possible score of 180, using the following steps:

1. Count up the total number of multiple choice questions that you answered correctly (out of a possible 70).

2. From the number correct, subtract $\frac{1}{4}$ of the total number answered incorrectly. (Questions left blank do not count against you here.)

3. After the subtraction, multiply the number obtained by 1.286 to weight the multiple choice total equally with the free response.

4. Add the total score from the free response questions to the number obtained in part 3.

5. The following scale loosely approximates the assignment of scores on actual AP exams (though this varies from year to year, depending on the difficulty of each exam).

Numerical Range	AP Score
124 – 180	5
71 – 123	4
59 – 70	3
30 – 58	2
0 – 29	1

These numerical ranges are only approximate indicators of a student's predicted success on the AP Physics B Examination. The actual examination will have carefully balanced percentages of test material between the multiple-choice and the free-response sections, based upon the syllabus printed earlier in this text.

PSAT SAT ACT*

Take Kaplan. Score higher.

Getting into college is more competitive than ever. A high SAT or ACT score can make all the difference on your college applications.

With more than 60 years of experience, Kaplan delivers a broad range of programs to help guide you through the college admissions process, including **SAT, PSAT and ACT prep classes, online courses, private tutoring and one-on-one college admissions consulting.**

For more information or to enroll, call or visit us online today!

1-800-KAP-TEST
kaptest.com

*Test names are registered trademarks of their respective owners.

The First Step to a Higher Score on Your Next AP* Exam!

Save $2 on the purchase of your next Kaplan AP guide♦

When you purchase one of the following titles:

- Kaplan AP Biology
- Kaplan AP Calculus AB: An Apex Learning Guide
- Kaplan AP Chemistry: An Apex Learning Guide
- Kaplan AP English Language & Composition: An Apex Learning Guide
- Kaplan AP English Literature & Composition
- Kaplan AP Macroeconomics/Microeconomics: An Apex Learning Guide
- Kaplan AP Statistics: An Apex Learning Guide
- Kaplan AP U.S. Government & Politics: An Apex Learning Guide
- Kaplan AP U.S. History: An Apex Learning Guide

Publisher's Mail-In Rebate
Simply complete the **publisher mail-in coupon in this book**, attach it to the **receipt from your new Kaplan AP guide** (please see eligible titles above) as proof of purchase, mail it to Simon & Schuster Customer Service, and we'll send you $2.

♦ Note: The rebate offer in this book is not valid for this book. The $2 mail-in publisher's rebate is valid for the purchase of a second Kaplan AP guide.

Please Send Me My $2 Kaplan AP Publisher's Rebate!

I am enclosing:

1. this completed mail-in publisher's coupon from my *Kaplan AP* Physics B: An Apex Learning Guide*
2. the sales receipt of my **new** Kaplan AP guide as proof of purchase

Name (to whom check should be made payable): _____

Address: _____

City/State/Zip: _____

Telephone: _____   Email address: _____

Mail this coupon to:
Simon & Schuster Customer Service
100 Front Street
Riverside, NJ 08075–1197

Offer valid from September 1, 2002 through February 28, 2005. Mail your claim within thirty days of purchase. No reproductions of cash register receipt or mail-in certificate accepted. Offer good in U.S. except where prohibited, taxed, or restricted by law. Not valid with any other offer. Limit one per household. Requests from groups, post office boxes, or organizations will not be honored. Manufacturer reserves the right to reject incomplete submissions or submissions otherwise in violation of these terms without notification. Manufacturer not responsible for late, lost, or misdirected mail. Please allow 6-8 weeks for delivery.

*AP is a registered trademark of the College Entrance Examination Board, which neither sponsors nor endorses this book or offer.